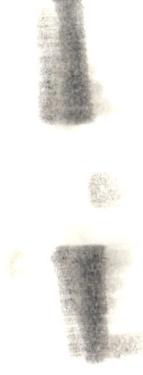

Global Radio

Global Radio

From Shortwave to Streaming

Shaheed Nick Mohammed

LEXINGTON BOOKS
Lanham • Boulder • New York • London

Published by Lexington Books
An imprint of The Rowman & Littlefield Publishing Group, Inc.
4501 Forbes Boulevard, Suite 200, Lanham, Maryland 20706
www.rowman.com

6 Tinworth Street, London SE11 5AL

British Library Cataloguing in Publication Information Available

Library of Congress Cataloging-in-Publication Data Available

ISBN 978-1-4985-9495-0 (cloth)
ISBN 978-1-4985-9496-7 (electronic)

In loving memory of my grandmother
Karriman "Baby" Mohammed
with whom I listened to the radio.

Contents

Preface

Technical details, policy measures and political debates are part of the history of global radio. While necessary, these discussions often fail to capture the fascination and romance of a medium that captivated the world, bringing distant voices and music to eager signal searchers who invested their time, money and expertise to experience the joy of "listening in." Before launching into the exciting and often surprising evolution of radio as a global phenomenon, we must also attend to some necessary details about the processes and procedures that have yielded the story to follow. Additionally, the present work would not be possible without the benefit of many kindnesses from several key individuals at libraries and research institutes that I visited over the course of my fieldwork.

Library staff at the Guyana National Library, the University of Guyana, the National Library of Trinidad and Tobago and the Alma Jordan Library at the University of the West Indies at St. Augustine in Trinidad and Tobago were all especially helpful. Similarly, the staff at the Sir Walter Rodney National Archives in Guyana went out of their way to accommodate my requests under difficult circumstances. My own library staff at the Altoona campus of the Pennsylvania State University were also especially helpful not only in locating materials but also in securing access to collections in foreign countries on my behalf.

METHODOLOGICAL NOTE

The current work combines archive research with mass media content as primary sources of data. The narrative relies on analysis of historical and policy documents but also uses mass media reports including archival newspapers, radio broadcasts and films as primary sources of data. Fundamental to this approach is the notion that mass media have demonstrated their value over many decades as a rich source of historical information. The approach also assumes that the media record is a relatively accurate reflection of the historical record. This is a process that has evolved among many researchers for several decades with a growing awareness that ubiquitous mass media have served (sometimes unintentionally) as a powerful historical record, particularly where several media have covered historical developments from multiple perspectives.

There is no effort to restrict data to only such sources, however, and the approach welcomes textual material from published books, particularly those of the time periods under investigation, papers and letters of key players, as well as both direct and indirect input from persons familiar with the issues being investigated. Thus, the approach includes conversations, both formal and casual, with audience members whose recollections of media content and experiences can lend context to the media reports and documented histories. In this approach, data thus includes media reports and information from sources such as accounts of parliamentary and congressional hearings, biographies and statutory reports. These are combined with interviews, both direct (where possible) and as recorded in texts and media reports with persons involved with the evolution of historical phenomena.

As modern mass media have evolved, so too have their forms. Some of these forms have become increasingly ephemeral. Web content, for example, changes rapidly and websites themselves may disappear without warning. For this reason, a wide range of digital archives for modern media have become indispensable for media-centered historiography. Several archival websites have captured snapshots of websites over the decades since the public availability of the World Wide Web, allowing glimpses into the past and examination of what might have been obscure or short-lived sites.

Additionally, while seminal texts on radio history have necessarily formed part of the research base, the focus on global reach (particularly with regard to small or remote geographical territories) has necessitated a fair amount of original material being collected from broadcasting publications, newspapers and other popular sources as well as station publications and documents. In some instances, letters to broadcasters and reception report cards have also proven useful.

FIELDWORK

Fieldwork for the present volume involved trips to locations including the Republic of Trinidad and Tobago in the Caribbean and the Cooperative Republic of Guyana in South America. Additionally, the work draws on earlier fieldwork exercises for other projects including trips to the Navajo Nation in Arizona.

In Trinidad and Tobago, primary sources included material at the Alma Jordan Library at the University of the West Indies in St. Augustine, Trinidad. At the West Indiana Collection, librarians made available numerous resources including the documents of radio pioneer Kamaluddin Mohammed as well as government policy papers and newspaper archives. The National Library and Information Services facility at Port-of-Spain was also an invaluable resource, providing, as it did, out-of-print brochures and other documents chronicling the history of radio in Trinidad and Tobago.

Three research sites in Guyana were important to the current work. The Sir Walter Rodney National Archives in Georgetown, the Guyana National Library at Georgetown and the University of Guyana Library at Turkeyen all provided important background material on the development of radio and its social and political uses in Guyana. Additionally, the opportunity to speak with the chief executive officer of the relatively newly constituted National Communications Network was instructive and gave some perspective, particularly on the approach to streaming and new technologies.

As Cambridge (2015) has noted, the condition of much of the material at the Sir Walter Rodney National Archives in Guyana is cause for concern. The historical newspaper collection is tremendously valuable but can scarcely be consulted anymore as the volumes are badly deteriorated, particularly around the time of the start of radio and radio-like services. Lack of digitization means that the original physical copies must be brought out to the reading room and handled. Several of the volumes may also be required in search of a particular story or group of stories since there is no index of articles in either physical or digital form. The delicate condition of the newspapers results in breakage and further deterioration, a fact well known to the staff members, who are often reluctant to allow access to these volumes.

Collections at the Guyana National Library were less ancient but well organized and useful. Their clipping files were well kept but only went back to the early 1970s. The University of Guyana facilities were also quite recent by comparison. Their collection of theses and dissertations regarding media in Guyana dated back to the 1970s as well, and though the collection was very useful, restrictive policies on copying made progress slow and tedious.

Introduction

One of the major failings of existing written histories of radio is their focus on the United States and Europe as the primary sites from which the medium emerged as a global force. They tend to make little mention of contributions from other parts of the world and how radio figured in the development of emerging nations. At the same time, it is difficult to conduct any responsible analysis of the emergence of global radio in its various forms without reference to the many important developments in the imperial centers of Europe and in the commercial market of the United States that drove the technology and its adoption. Thus, it remains necessary to examine the traditional accounts of radio in Europe and the United States to contextualize and extend the discourse into the experiences of other territories, including many that would not develop their own radio until many years after the emergence of radio as a global force. The present work attempts to balance the influences of the pioneers in global centers with the impact of those on the periphery who sought engagement with the distance-bridging and immediate phenomenon of radio by which their world might be opened with the turn of a knob.

What lies hidden in all of this is a rich and complex tapestry of interwoven influences in which radio has played meaningful roles throughout the development of modern society in many countries. The present volume examines the evolution of the global dimensions of radio from its early days of terrestrial/atmospheric transmissions to its modern manifestations across Internet-networked devices and its integration with mobile and social platforms. This study also examines the roles, functions and implications of global radio within and among former colonies and other developing nations, many of whom have invested in radio as a source of national development and identity. Some of these nations started their experiences with radio as colonial outposts for whom audio broadcasting via radio waves may have been little more than a nod to keeping colonial settlers informed of developments in the imperial centers. From such beginnings, many of them have become centers of broadcasting with regional and international reach and impact. We shall explore several examples of that evolution in the context of radio and its global scope.

Several authoritative histories already exist of so-called "international radio broadcasters," government-backed operations engaged in the specific business of reaching out to foreign, global audiences. Among these,

Graves' (1941) early outline of cross-border broadcasting and Wood's (1992, 2000) much more comprehensive and detailed two-volume study stand out. There are also several excellent regional studies of radio including, notably, Boyd's (1999a) study of electronic media in the Arab world. However, such studies have demonstrated several blind spots, particularly regarding the extent to which otherwise "local" radio broadcasters enjoyed regional and global reach and influence. The present work is not meant to be a substitute for an analysis of formal global broadcasters, but rather to show how several technologies and approaches—including the international radio broadcasters—have contributed to radio as a global force.

Existing studies, partly due to their age, also fail to address more recent fundamental changes in media technologies, audience involvements and media choices that now impact radio. These emerging media realities also now cast some doubt on the established notions of what constituted global or international broadcasters. With the ability to break out of the constraints of a system of shortwaves and relays that became formalized into the mouthpieces of predominantly powerful Western nations (and some powerful Eastern players as well), it becomes possible to envision the globalization of radio in diverse ways. The following pages examine several dimensions of radio's global reach including formal international broadcasters as well as the global operations of local stations (many from developing nations) both modern and antique.

In so doing, the narrative tracks the global trajectories of modems stations (linked by digital networked technologies to distant audiences), but also recounts the global exploits of some of the pioneering efforts at audio broadcasting via radio waves with stations such as KDKA in Pittsburgh and WGY in Schenectady creating their own global stir with enthusiastic listeners from Cape Town, South Africa, to the jungles of British Guiana. This early global radio quickly moved beyond the occasional brief reception of a stray signal into a formalized system of relays using powerful antennas to pick up distant signals which were then relayed to various local audiences such that the United Kingdom eagerly received concerts and talks from the United States while British colony residents in South America followed news from London and music from the Netherlands. Streaming technologies may have brought radio's global dimension to the forefront once again today, but almost from its start, global reach was an important part of the wireless audio dissemination systems (and their social effects) we know as broadcast radio.

The present work contextualizes radio as a part of "globalization" and attempts to 1) cast a sharper focus on the global dimensions of early radio, including the global reach of major US and European pioneers as well as the global practices of small and often underpowered stations in the "developing world," 2) outline several of radio's many roles and functions in global social and political issues including wartime propaganda, colonial cohesion and (in several cases) the transition into postcolonial nationhood and 3) examine the changing landscape of radio, including small developing-nation, ethnic and minority stations who now reach global audiences through digital networks. Underlying all of this is a view of audio radio broadcasting as a persistently global medium, enabling instantaneous and global communications in various forms from its beginnings to its present manifestations.

GEOPOLITICAL CONSIDERATIONS

Much of the content in this volume deals with relationships among nations and attendant issues such as conflict, war, colonialism and power relationships. The present work proceeds with due attention to the sensitivity of such topics as the role of radio in the geopolitical wrangling and open warfare that manifest itself as World War II. The judgements of history regarding the dangers of propaganda and suppression of free expression are evident in this narrative. So too does this work take seriously the dangers of hate speech in broadcasting.

The term "propaganda" itself deserves some attention, having been used in differing senses over the history of radio. The pejorative sense of information meant to strategically misinform, particularly in combat or conflict situations, had already emerged by the time of Lasswell's (1927) study in which the word "propaganda" had "come to have an ominous clang in many minds" (p. 2), particularly since "people were everywhere warned during the war to beware the noxious fumes of enemy propaganda" (p. 3).

Yet, words evolve their meanings slowly and concurrent variants are not uncommon. Thus, as Wood (1992) has noted, propaganda in its broader sense of message dissemination continued to be used in situations such as debates over advertising on the radio in the late 1920s and early 1930s. Whereas most times the term is encountered in the present work, it is in keeping with the modern, relatively common reference to strategic misinformation, there will be instances where its older and broader sense of simply spreading messages will often be relevant.

Despite the incompatibility of information control with modern Western notions of freedom of speech, the current narrative also recognizes the extent to which notions of self-reliance and nationalism have been important to various emerging nations. Many so-called "Third World" or "developing" nations have evolved from histories of colonial oppression and domination. Those experiences prompt suspicion of policies that appear to favor rich and powerful nations. These more recently independent nations are often wary of the possible effects of media content from former colonial masters or new international economic powers. As the present work examines radio in the interplay of nations, these several considerations emerge at various points. It is often difficult to avoid the biases of the modern, dominant, liberal free-speech paradigm, but the media records and other data often demonstrate the tensions between efforts to spread information and efforts to control its spread.

RADIO AS SOCIAL PHENOMENON

The current volume maps radio as a global event over more than a century. For all the numerous and diverse technical developments covered here, the social uses of radio are much more important. As several authors have noted prior to this work, radio is much more than the sum of its technical parts. The ability to take the voice and music over great distances and to instantaneously capture the attention of mass audiences was not simply a technical achievement—it also marked the inception of mass audiences and mass society.

Taken in a global perspective, radio's mass audiences marked the start of humankind's greatest age of shared knowledge and information. The contentious politics of wartime propaganda, the shared enjoyment of entertainment across national boundaries and even the cultural continuity of ethnic broadcasting are all part of global radio and its power. Over time, global reach has become easier, but global audiences have also become more fragmented. While radio has yielded to television and the Internet, it has also harnessed digital technologies to reach its fragmented and far-flung audiences globally, reinforcing cultural ties for diaspora societies and displaced listeners wherever they might be.

ONE

Globalization, Radio and Culture

The end of the twentieth century ushered in an age of fascination with the potentials of global connections particularly focused on the ability of networked digital technologies to reduce or eliminate both distance and borders. Authors such as Frances Cairncross (1997) proclaimed "The Death of Distance" arising from these technologies, and neo-liberal economists and academics alike were increasingly likely to associate digital networked technologies with the emergence of global markets and attendant social and cultural shifts generally termed "globalization."

According to Rzpeka (2011), the term "globalization" first emerged in modern usage in 1959 with the *Economist* magazine. Webster's dictionary included it in 1961, and while its initial use had to do with the global trade in automobiles, the notion of globalization emerged into popular thought with the work of communications guru Marshal McLuhan in the 1960s and 1970s. Of course, at the time, McLuhan was not referring to the Internet or other digital networked technologies but rather radio and television (already reinforced with the global reach of satellite technology).

Though there are clear and demonstrable changes in global communications and commerce, there is less clarity on what exactly constitutes "globalization." The term is an elusive (if ubiquitous) one, involving assumptions about the roles and functions of technology as well as concerns about cultural flows from powerful nations and the impacts of such flows on receiving nations.

Among the many criticisms of globalization is that it is a short-sighted concept, rooted in an emphasis on recent technologies and ignoring the powerful and enduring influences of global commerce and culture that have been forged over centuries of global interactions among nations (Mohammed, 2011). In this regard some writers have tried to expand the chronological scale of what we now call globalization to specific periods, citing, for example, Wallerstein's (1974) notion of the emergence of the so-called "World System" manifest in European colonial expansion during the eighteenth and nineteenth centuries or even more recently to Machlup's (1962) notion of the emergence of an Information Society during the 1960s.

Scholars have differed on the extent to which even the modern technologies of globalization should be considered sufficient tools of social or economic improvement as some techno-determinists (Negroponte, 1995; Rhinegold, 1993) appear to suggest. Staley (1998) was among those who warned that not only did disparities in access to the technologies of globalization perpetuate disparities among groups and nations but also that the prescriptions of free markets and industrialization reflected the values of one set of nations that could be seen as being imposed upon others.

Given this wide range of relevant fields in which globalization remains a key issue, investigations and analyses of globalization have emerged from a wide range of perspectives, including media, political economy and cultural studies. Globalization (however defined) is important in the present analysis of radio technologies in several ways. At the outset, the emergence of radio and its ability to instantaneously traverse increasingly great distances began to suggest a global scope. As radio crossed borders and created simultaneous audiences in different regions and nations, it began to create the conditions for global audiences. As we shall examine, global audiences for radio would emerge and change as hostilities led to war and as global broadcasters began to take form. These global audiences might be sympathizers of one political cause or another, enemies of the broadcasters or members of colonial communities awaiting news from the seat of the empire.

As international radio became increasingly specialized into state and religious broadcasting, local radio began to emerge with stronger AM and FM services to small local audiences. Eventually, however, the allure of the Internet and the World Wide Web began to entice otherwise local broadcasters with the potential to reach the entire globe. The reach to which early radio practitioners aspired eventually became available to broadcasters almost a hundred years later, this time without the problems of reception and fading. In these and other ways, broadcast radio has been involved in the mediated processes of globalization that were new in the early twentieth century and reinvented in the twenty-first. Radio's role in the global process of conflict, influence and information in the intervening period has formed an essential part of the globalization dynamics that are still debated today. Its roles and functions in maintaining empire and mitigating propaganda demonstrate the diversity of globalization's many facets and the integral role of communications media, with radio being one of the key conduits because of its affordability, ubiquity and ease of use.

RADIO, CULTURE AND CHANGE

Radio, as Hilmes (1997) has powerfully argued, is not merely an agglomeration of wires and tubes or microchips. It is more than a set of stations involved in broadcasting messages that span local communities as easily as it spans oceans or crosses national borders. For Hilmes, as for other scholars of media and society, radio and other media are far more than either their technical embodiments or their industry organizations would suggest. Radio, like other media, are what cultural studies scholars like Stuart Hall (1989, 1997) have termed "signifying practices" because they are involved in process of meaning and attendant cultural flows. Radio, a as global phenomenon and as a signifying practice, is therefore important as a contributor to global culture. The cheapest and most accessible of all mass media, radio has enabled global influences across cultural boundaries and enabled cultural flows within cultural boundaries. Radio, in its local and its global dimensions, has been an integral part of globalization long before the term was even in vogue.

Outside of the more subtle influences on culture that might be inferred from the dominance of American music on stations around the world or the continuing influence of the British Broadcasting Corporation (BBC) in former British colonies, radio has been much more directly used as a deliberate tool of public education and behavior change in several parts of the world. Perhaps no such efforts have been better documented than the so-called "entertainment-education" or "enter-educate" campaigns for public education in past decades.

To tackle the rapid spread of HIV/AIDS in parts of Africa and the swirling misinformation that hampered efforts at prevention, governments, international agencies and NGOs employed radio as a tool in social transformation. In Tanzania, the government, working with its Population and Family Life Education Program (POFLEP), the United Nations Fund for Population Activities and researchers from the University of New Mexico, launched a radio soap opera in 1993 called *Twende Na Wakati* (Let's Go with the Times) that ran for several years and reached a majority of Tanzania's population. The project used local actors and regional scriptwriters to create a dramatic narrative in the native language that followed the exploits of a truck driver (one of the professions that acted as a vector for the disease at the time) and his experiences on the road and with his family. In that radio drama, listeners (often listening as groups with community radio sets) vicariously followed his experience and that of his wife as they faced the threat of the disease and discussed its implications, proper prevention methods and the myths and misinformation surrounding it. Research conducted during and after the series showed that listeners not only learned more than non-listeners about HIV and prevention of the disease but were also more likely to discuss these matters with others (Mohammed, 2001; Singhal & Rogers, 1999; Svenkerud, Rao & Rogers, 1998).

Efforts to influence perceptions of family size, the acceptability of family planning and the dangers of HIV in the (heavily Roman Catholic) island nation of St. Lucia bore the stamp of radio as well. During the 1990s, a radio soap opera series entitled *Apwe Plezi* (After the Pleasure), featuring local voices and writing from local and regional scriptwriters, addressed issues of family planning and contraception, resulting in measureable changes in the community's perceptions of ideal family size, contraceptive use and the dangers of HIV (Vaughan, Regis & St. Catherine, 2000). In India, a 1990s radio soap opera series entitled *Tinka Tinka Sukh* (Happiness Lies in Small Things) served to influence deeply held social beliefs about dowry payments detrimental to women's rights (Singhal & Rogers, 1999). Other smaller projects in Jamaica, Trinidad and Tobago and other Caribbean islands have used radio to promote social information and behavior change on issues ranging from nutrition (Food and Agricultural Organization of the United Nations, 1997) to childcare, migration and teen pregnancy (Caribbean Family Planning Affiliation, 1995). While such concerted radio campaigns have faced criticism for the involvement of outside agencies and governments and the imposition of foreign values on local populations, there is little question that they have been proven effective at message dissemination and cultural change (Singhal & Rogers, 1999).

Radio has also been a tool for cultural conservation. The Hopi Native American reservation occupies about 2,500 square miles in Arizona with a population of about 7,000. As in the rest of the United States, mainstream media is the dominant linguistic influence with English-language programming on satellite television and commercial radio stations reaching these areas. Dukepoo (2013, p. 22) described the influence of Native American radio station KUYI 88.1 FM among its Hopi audience in and around the small reservation village of Keams Canyon, Arizona, since 2000, particularly with regard to its role in preserving their traditional language in the face of linguistic decline:

> Today, inside almost every home, office, and vehicle, the radio is tuned to KUYI, which is also Hopi for "water." The station has become an integral part of life on the Hopi reservation, as listeners start their day with a morning greeting in the Hopi language. . . . KUYI is fortunate to have a staff of volunteer DJs who are fluent in Hopi. They share the Hopi language through the airwaves, ensuring the language is heard and spoken.

Whereas in past decades from the inception of Native American radio in the early 1970s, stations such as KUYI and its counterparts like the Navajo Nation's *Diné* language KTNN at Window Rock, Arizona, served as tools of cultural maintenance only for the members of their contiguous listening audience, their scope of influence has, today, expanded to all members of their linguistic group no matter their location. As Dukepoo (2013, p. 22) has noted, KUYI now spreads its linguistic and cultural influence globally since "the sharing of language creates a strong connection with 'home'" among Hopi listeners "who live around the world and listen online."

In other global contexts, radio has emerged as a vehicle for traditional cultures and the cultural imaginings of diaspora populations. As we shall examine in greater detail in a later chapter, descendants of Indian indentured laborers in Trinidad and Tobago found a haven for their cultural expression on commercial radio stations dedicated to Indian music and cultural content starting in the early 1990s. Even though these diasporic Indians had, by that time, lost most of their ancestral languages, these radio stations served as a rallying point for their community.

ORIGINS OF RADIO AND ITS EARLY USES

Numerous actors and corporations were behind rapid developments in the communications technologies known as "wireless" in and around the early 1900s. A flood of proprietary techniques and mechanisms both pushed the medium forward and held it back. British officials hoping to establish a great imperial communications system lamented, for example, that competing technologies, mechanisms and methods of multiple industrial interests prevented a true unified system of wireless communication.

The profusion of names, credits and claims hails back to much earlier developments in the science of broadcasting and breakthroughs upon which radio was built. Garratt (1994, p. 5), for example, argued that Michael Faraday's work laid the foundation for radio communication, arguing that "without Faraday there would have been no Clerk Maxwell, no Hertz, no Marconi, and it is almost certain that the advent of wireless telegraphy would have been delayed by many years." Yet, even so, Garratt related that Faraday depended on the earlier work of the Danish scientist Hans Christian Ørsted (sometimes rendered as Oersted), who experimented with magnetic fields and posited the link between electricity and magnetism. The US Federal Communications Commission recognized the diverse genealogies of radio broadcasting, making reference to radio's "multiple parentage" and the numerous individuals credited with "fathering" the medium (Federal Communications Commission, 2003–2004, p. 1), including among them Henirich Hertz, Nikola Tesla, Ernst Alexanderson, Reginald Fessenden, Edwin Armstrong, Guglielmo Marconi and Lee de Forest.

DIVERSE FORMS

At different times in its evolution, what we know of today as radio took various forms, ranging from wireless telegraphy to point-to-point voice communication and, eventually, audio broadcasting or "broadcast radio." While each of these forms depended on the same principles of electronic signal propagation, they differed in their technical manifestations and their uses. Indeed, the mass medium of radio was a fairly late addition to the range of wireless technologies that evolved in the late 1800s and early 1900s. On this point Satia (2010, p. 829) has called "strange" the fact that "public broadcasting" or creation of a mass medium was not among the early emphases of wireless radio, noting that "scientists and the press" of the time in fact bemoaned the technology's "ability to transmit to multiple receivers." This gradual evolution into a medium of mass communication began with the development of the ability to harness and control electrical signals through the air or what is often simply referred to as "wireless."

Wireless

Advances in wireless message transmission emerged in the context of scientific inquiry into electricity and its use in the propagation of signals alongside an already thriving industry in wired communication including the telegraph and the telephone. The telegraph and telephone had also, in the course of their development, demonstrated international communications capacity, particularly when interlinked by undersea cables. The use of underwater cables to bridge telegraph transmissions over local bodies of water had been demonstrated since 1842 and England and France created an international undersea telegraph cable link in 1851 (Wheen, 2011, p. 19), but the true global goal was to link the United Kingdom and the United States. Technical and financial challenges delayed this development until 1866 when messages previously requiring the duration of a ship's voyage across the Atlantic Ocean could be transmitted in seconds instead. Yet, despite the promise and confirmed performance of wired systems during the 1800s, the lure of transmission without wires remained strong.

By the end of the nineteenth century, scientists and experimenters were familiar with electric wave propagation through the air and methods of receiving them, though no one (until the youthful Marconi) had been able to combine and improve these concepts into a workable system of message exchange "available for use at considerable distances" (Towers, 1917, p. 199). Following on the work of Oersted, Maxwell, and Hertz, Guglielmo Marconi (born April 25, 1874) was able to create an electrical disturbance at one point and receive a signal corresponding to that disturbance through the air at some distance away. Between 1894 and 1895,

Marconi, working on his family's property in Bologna, Italy, succeeded in sending a signal to a receiver, first just across the garden and then (by some reports) up to some two kilometers away.

Marconi was born of an Italian father and a British mother. Having failed to impress authorities in his native Italy of the usefulness of his discovery (Adams, 2012; Migliore & DiPierro, 1999), Marconi, with the help of his family connections, travelled to England and approached the British War Office, pitching his remote electrical signal transmission not as a form of communication, but rather as a method of remote control for boats or torpedoes (though from the evidence of his patent applications, Marconi did refer to his techniques as improvements to telegraphy). Adams (2012, p. 44) has also noted the role of Marconi's British mother in advancing his ideas and family connections that allowed him "access to highly placed officials such as Director Preece of the British Post Office, his patent attorneys, and anyone else who could elevate his ideas and get them noticed."

Marconi was clearly not the first to conceive of the possibilities of information transmission without wired connections. Fahie (1901) traced this distinction to the work of a Spanish physician named Francisco Salvá y Campillo (1751–1828) who, in 1795, described a system for transmitting a signal over a long distance using a body of water as the means of conduction. Lindell (2006) credited Samuel Morse (1791–1872) with the conceptualization and development of wireless telegraphy over water and conducting the first experiments in this regard around 1842.

With the emergence of distance-bridging techniques such as the wired telegraph, British authorities (both at home and in their colonies) vested much of the control and policy decisions regarding emerging communications technologies in the British Post Office. As Garratt (1994, p. 81) has chronicled, it was the British Post Office authorities who controlled and observed the demonstrations of Marconi's wireless signaling devices in 1896 and 1897. Representatives of the War Office were also present.

Reporting on the Marconi experiments, the *New York Times* of July 19, 1987 ("Topics of the Times," p. 4), described the system (somewhat derisively) as a "contrivance for wireless telegraphy" consisting of "an electrical machine on a stick with an electric wire wrapped around it" by which "electric vibrations" were "transmitted to another stick attached to a receiver." Yet, by August 14th of the same year, a small article in the *New York Times* ("Nuggets," 1897, p. 4) encouraged that the wireless telegraph be tried in Alaska where it might be most useful. Interest was such that the papers started reporting incremental achievements such as an experiment in Berlin in October of 1897 that achieved transmission of signals wirelessly over 21 kilometers ("Telegraphy Without Wires," 1897, p. 7).

Fahie (1901, p. 135) noted that the British Post Office's Engineer-in-Chief Sir William Preece had "made the subject of wireless telegraphy a special study for many years, his first experiment dating back to 1882″ (when Marconi was eight years old). The Marconi experiments were conducted at the site of earlier trials in which William Preece had demonstrated the possibilities of inducing a signal in one circuit from another unconnected electrical circuit using parallel lengths of cable.

Independent of Marconi and the Post Office, elsewhere in the British Empire, others were advancing work in areas such as remote signal transmission. In Calcutta, in 1895, Indian scientist Jagadish Chandra Bose (1858–1937) demonstrated a system for transmitting electrical signals wirelessly, even through obstacles such as walls, over short distances using 5 mm waves (Mitra, 1997; Mukherjee & Sen, 2007). A Bose biography (G. A. Natesan & Co., 1921, p. 9) claimed that Bose was the first to achieve remote transmission in 1895 at the Calcutta Town Hall where "he transmitted ether wave through a solid wall and a line of men and made it displace a heavy weight, ring a bell and explode a miniature mine placed in a closed room." Bondyopadhyay (1998, p. 259) has, in fact, argued that the "mercury coherer" involved in Marconi's successful transatlantic transmissions (from Cornwall, England, to Newfoundland in Canada in December 1901) was a Bose invention first reported to the April 27, 1899, meeting of the Royal Society of London.

This was not the only challenge to Marconi's claims about wireless telegraphy. The *New York Times* reported on January 2nd of 1898 on the competing claim "that the real inventor of wireless telegraphy was a Frenchman named Bourbouze" in 1871:

> In November the wires leading to Paris had all been cut. It was then that M. Bourbouze conceived the idea of communicating by electricity and without wires. He experimented for many hours a day in his back yard . . . and on the night of Jan. 10–11, 1871, a dispatch was sent along the Seine from the Pont National to Saint-Denis, a distance of nearly twenty-five miles. ("In Foreign Lands," p. 19)

Fahie (1901, p. 99) suggested that inventors who had already patented some achievements in wireless telegraphy felt cheated over Marconi's fame, as was the case with American physicist Amos Emerson Dolbear (1837–1910), whose friends in America claimed for him "the discovery of the art of wireless telegraphy *a la Marconi*" since Marconi's apparatuses and methods so resembled work Dolbear patented in 1882.

Later, in perhaps the most famous of these contentions (after Marconi and his company were comfortably established as household names), inventor Nikolai Tesla sued Marconi over the patent rights to wireless telegraphy. The Nikolai Tesla Company filed the case in the US district court in New York, naming the defendant as the Marconi Wireless Telegraph Company of America. This suit hinged on several of Tesla's variously substantiated claims of having developed the ideas that Marconi touted as his own. Molisch (2011, p. 4) has argued that "while Tesla was the first to succeed in this important endeavor, Marconi had the better public relations."

Marconi's public relations had, by that time, included several high-profile events across the globe. In July of 1898 Marconi capitalized on an invitation to cover the Kingstown Regatta in Ireland for the *Dublin Daily Express* (Regal, 2005, p. 26). As Moffett (1899, p. 100) described it, the paper arranged for a sending station to be placed on an observer vessel with a transmitting antenna some 80 feet up its mast sending signals to a receiving station at Kingstown from where they were telephoned to Dublin "so that the 'Express' was able to print full accounts of the races almost before they were over, and while the yachts were out far beyond the range of any telescope." Regal (2005, p. 26) suggested that Marconi's coverage of the Kingstown Regatta not only demonstrated the ship-to-shore feasibility of wireless telegraphy but also served to promote the man and his work, leading to a more international scope for his activities when James Gordon Bennett ("publisher of the New York Herald and a yachting enthusiast") invited Marconi (for the sum of $5,000) to employ wireless telegraphy to cover the America's Cup yacht race in New York in October 1899.

Despite the excitement among the yachting crowds, this achievement did not impress everyone. The journal *Forest and Stream* ("The America Cup," 1899, p. 336) had a less than laudatory mention of Marconi's technologies, noting that for a price "one could boast that he had been on the same steamer as Mr. Marconi, the wireless telegraphy expert" and "could send some silly message by wireless telegraphy to friends ashore and have it printed next day with his name in a morning paper."

The seafaring capabilities of Marconi's wireless also drew attention beyond that of the curious public. Marconi's success at broadcasting news and information at the America's Cup quickly led to interest from the US military, specifically the US Navy, who requested the equipment for testing "to determine the practicability of the system for short-distance signaling between squadrons at sea" (Fahie, 1901, p. 240).

According to Aitken (1985, p. 197), "what differentiated Marconi from his contemporary rivals was not his scientific knowledge, nor, initially, the distinctive excellence of his technology" but "his sense of the market, of where a demand for this new technology existed or could be created." Toscano (2012, p. 60) described Marconi's self-promotion as a kind of public relations campaign which included presentations to the technical and scientific community in which he represented the wireless as "human advancement." The business community and investors were also being drawn into the emerging success of wireless. The periodical *The Advance* suggested in 1903, for example, that great fortunes were possible from investing in the Marconi wireless system and made parallels with Bell's telephone, which was initially considered a novelty but became immensely profitable when the public began to think of it as a necessity, arguing that the "faithful friends" would eventually "reap the fortunes that are bound to follow the general adoption of the wireless system" ("Marconi Wireless: Great Fortunes," 1903, p. 258).

Marconi convinced important figures like Queen Victoria, King Edward VII and Theodore Roosevelt to exchange messages over his wireless system and enjoyed publicity from newspaper and magazine coverage of these events. The *Chicago Daily Tribune* in an article datelined Wellfleet, Massachusetts, January 19, 1903, reported that Marconi was exultant over his achievement of a 3,000-mile transmission "when he transmitted a message of greeting from President Roosevelt to King Edward direct from his station here to the one at Poldhu, Cornwall, England" ("Marconi Scores New Triumphs," 1903, p. 5).

Partly because of Marconi's publicity efforts, wireless telegraphy and the ability to transmit messages over distance without wires began to captivate the public imagination throughout Great Britain and elsewhere (Fahie, 1901). In the United States, despite Marconi's successes and popularity, public sentiment did not always support the continued domination of wireless from a foreign source, whether the Italian or his British company. In early 1920, as the industry anticipated the return of several wireless stations to private hands after the government takeover during World War I, newspapers announced the formation of the Radio Corporation of America (RCA) emphasizing both the global scope of the new enterprise and its American ownership. The *New York Times* of January 5, 1920, in reporting on the new twenty-million-dollar company, proclaimed "American radio to span the globe," describing the RCA scheme as an "American and international wireless system that is to link the countries of the world in exchanging commercial messages" and specifically noting that "the former British interests in the Marconi Company have been acquired by the General Electric Company, and the ownership and control of all operations of the Corporation are vested exclusively in Americans" ("American Radio," 1920, p. 1).

Voice over Wireless

Despite its broadening range of applications, wireless remained, for many years, an extension of the telegraph both in concept and in use, focusing on transmission of simple messages comprising dots and dashes. With proper arrangements in place, these messages could facilitate financial transactions and other relatively sophisticated actions. However, even the simplest messages involved a slow and cumbersome process in which trained operators exchanged Morse code dots and dashes to represent letters and numbers.

The evolution of Morse code wireless telegraphy into a medium for sound transmission involved various experimental forays. Rural Kentucky farmer Nathan Stubblefield, for example, used a version of earth-induction telegraphy to enable wireless voice communication among neighbors around 1902 (Lochte, 2001), which he later patented as a wireless telephone system. At around the same time, others also pursued the transformation of wireless telegraphy into sound transmission. The key to this advance was a move from the intermittent transmission of dots and dashes to the use of continuous wave technology. The ability to transmit voice and music would require equipment and techniques that enabled two separate wave disturbances to be transmitted: 1) a modulation wave that represented an electrical rendering of the sound at the source that would be reproduced through an identical modulation at the destination receiver and 2) a carrier wave to transmit the modulation signal over long distances.

Lee de Forest, Ernst Alexanderson and Reginald A. Fessenden were among the figures most widely credited with advancing wireless telegraphy into a form that would allow later development of audio broadcasting over radio waves. However, some sources also include a Brazilian priest named Roberto Landell de Moura with these inventors since he had some success with voice transmissions in the 1890s but was ordered to stop by his bishop, "who claimed that this device was witchcraft and transmitted the voice of the Devil" (Lochte, 2000, p. 96).

Around 1906 Lee de Forest (born August 26, 1873, in Council Bluffs, Iowa), concerned with improving how well receivers could pick up and amplify weak or distant signals, developed a vacuum tube called the Audion which evolved into the vacuum triode. While de Forest was experimenting with improving reception, Reginald A. Fessenden (born in 1866 in Quebec, Canada) was developing the ability to transmit actual sound from one point to another wirelessly. His success would depend, in part, on de Forest's advancements in reception and amplitude modulation as well as Alexanderson's development of continuous wave technology.

Ernst Alexanderson (born in 1878 in Uppsala, Sweden) migrated to the United States in 1901 and later worked for General Electric (Brittain, 1992). He started at General Electric in 1902 and eventually worked with Reginald A. Fessenden ("Equipment & Engineering," 1975). Helen May Fessenden (1940) recorded her husband's first success at transmitting voices wirelessly in 1900 at Cobb Island while he was working on developing a system of weather reporting for the US government. Helen Fessenden (1940, p. 81) reported that Reginald Fessenden and his associates first conducted audio tests in December of 1900 when "intelligible speech by electromagnetic waves was for the first time transmitted" though it was "poor in quality, since it was accompanied by an extremely loud and disagreeable noise due to the irregularity of the spark."

After several years of experimentation Fessenden made a public demonstration of wireless sound transmission in 1906. Helen May Fessenden (1940) recorded that Reginald A. Fessenden distributed special wireless audio receivers to ships of the US Navy and the United Fruit Company and later sent messages to the ships to expect a program of music and speech on Christmas Eve and New Year's Eve. She described the event in Reginald Fessenden's words as follows:

> The program on Christmas Eve was as follows: first a short speech by me saying what we were going to do, then some phonograph music. The music on the phonograph being Handel's "Largo." Then came a violin solo by me, being a composition of Gounod called "O, Holy Night," and ending up with the words "Adore and be still" of which I sang one verse, in addition to playing on the violin, though the singing of course was not very good. Then came the Bible text, "Glory to God in the highest and on earth peace to men of good will." (1940, p. 153)

Amateur Operators, Two-Way Radio and Experimentation

These emerging wireless audio transmission technologies did not automatically lead to broadcast audio via radio. For several years experimenters and hobbyists dabbled with these technologies, variously combined as wireless forms of telegraph, telephone and radio, testing their boundaries and their potential uses. Voice at a distance was becoming increasingly important and by 1915 engineers at the American Telephone and Telegraph Company (AT&T) demonstrated not just transatlantic but also transpacific capability (Austin, 1922).

As Schubert (1928, p. 195) has pointed out, what was then called "the radio art" was quite new, characterized as a "'fiddley' proposition, a thing of adjustments and tinkerings, of finding a good 'spot' on a crystal detector, of intricate 'tuning' by means of long coils that slid one within the other." According to Schubert (1928, p. 194), there were some five thousand licensed radio amateurs throughout the United States prior to World War I, "most of them youngsters who had tinkered with 'spark' radio." When the United States became involved in World War I, however, the government moved to shut down or take over most private broadcasting operations. According to Arceneaux (2012, p. 2), the ban, issued on April 6, 1917 (and lasting until mid-1919), was complete and strict with only a few exceptions made for operators such as those who were working on experimental transmissions for the navy such that "transmitting wireless signals was no longer permissible nor was the simple act of putting on a pair of headphones and listening to the airwaves."

Experimenters did not always focus on audio broadcasting and sometimes ventured into areas such as sending facsimiles and photographs using radio waves. Inventor C. Francis Jenkins, for example, claimed to have "produced the first photographs by radio, and mechanism for viewing distant scenes by radio" (1925, p. 4), and argued in 1925 that "the rapid development of apparatus for the transmission of photographs by wire and by radio may now be confidently expected" (p. 5). *Scientific*

American reported in 1924 ("Radio Notes: Short-wave," p. 357) on Nikolai Tesla's plans to transmit power to homes using radio in order to minimize the losses over distance inherent in wired power distribution. In 1926, Professor S. E. Dibble of the Carnegie Institute of Technology even predicted that "Radiocasting of Heat for Homes" was a matter of a few years away ("Radiocasting of Heat," p. 3).

More commonly, experimentation with radio involved existing technologies, notably the telegraph and the telephone, with emphasis on integration with these established systems. A 1921 text on radio, for example, made reference to long-distance communication using audio as "radio-telephone" while noting its distinct advantage over the "radio-telegraph" of not needing a specialist Morse code operator and its use "in places over land and over water where it would be either impossible or extremely uneconomical to use wires" (Morecroft, 1921, p. 646).

Stone (1919) noted that despite the rapid and widespread adoption of the radiophone it was not expected to eclipse the wire telephone for general use due to the problems of interference that would doubtless emerge. However, Stone (1919, pp. 293–94) also noted that the two-way system of radiophone communication was well suited to mobile applications such as was already the practice at sea and in the air so that "radio communication between units of a fleet and between individual ships and aircraft within the various units" was already "absolutely essential for the proper handling of the modern fleet" and was being used to enable communications on land among other interests such as police departments in New York and Chicago.

Following the development of sound capability, the pre-existing preoccupation with increasing the distance of transmission dominated discourse and experimentation surrounding radio. The quest for sound transmissions over increasing distances spread throughout the world. The English-language *China Press* newspaper reported on March 4, 1925 ("Shanghai and Osaka Talk," p. 1), that two-way radio communication was successfully established between Shangai and Osaka. In November of 1925, amateur radio operators in the United States and South Africa similarly boasted of establishing wireless contact ("Amateur Radio," 1925). Early newspaper reports in Guyana also indicated that outside of efforts to develop broadcast radio for entertainment, enterprising mining interests were using radio technology to maintain communications between mining bases and operations in the nation's vast interior as early as the 1920s and 1930s ("Men Going," 1927).

BROADCAST RADIO (AUDIO BROADCASTING)

As the "radio arts" evolved, wireless telegraphy and point-to-point wireless audio diverged from "broadcast radio" with its dedicated content providers, simpler receiving systems and broader public participation. For Brown (2005, p. 28), those involved in emerging systems of broadcast radio regarded the earlier point-to-point wireless telegraphy as part of an old order, to be eclipsed by broadcast radio. Whatever the strength of that argument, the new broadcast radio phenomenon was, however, by no means a foregone conclusion as it emerged in the early 1920s.

Stone (1919, p. 296) described "radiophone broadcasting" as emerging from the work of enthusiasts, experimenters and equipment manufacturers and retailers and being used for "the public broadcasting of music, press news, financial and crop reports, weather reports and storm warnings" but with ever-increasing diversity of content:

> Briefly, radiophone broadcasting consists of the operation of radio telephone transmitting stations upon fixed schedules and wave lengths . . . with definite programs arranged for the entertainment and instruction of thousands of private receiving set owners. Grand opera concerts, musicales, church services—with sacred music and sermons—and complete press reports of important sporting events, the news of the day, etc., are sent out over the radiophone and received by stations within radii, in many cases, of 2,000 miles.

Though the price and complexity of receiving equipment were already on the decline in the period leading up to the 1920s (a good receiver being similar in price to a good phonograph at the time), several elements of radio broadcasting were still quite novel, including the idea of connecting a loudspeaker to the receiving equipment so that more than one person could listen at the same time (Stone, 1919).

This practice of disseminating information over radio waves was termed radiocasting or broadcasting though not even the terminology was initially certain. In June 1924 at the annual convention of the Associated Manufacturers of Electrical Supplies in Atlantic City, New Jersey, their radio division voted to officially abandon the term "broadcasting" in favor of "radiocasting." According to a newspaper report of the time ("'Radiocasting' Is Adopted," 1924, p. E7):

> "Radiocasting" will henceforth be the term applied to indicate the spreading of sound through the air. . . . The radio supply manufacturers made the change on a recommendation of a committee which reported that the word "broadcasting," according to the dictionary, "has to do with the sowing of seed of material substance."

Clearly, however, "radiocasting" did not win public support as the preferred term for audio by wireless though several competing terms such as "radio telephony" and "wireless telephony" persisted for several years until the widespread adoption of radio broadcasting and reception technologies resulted in the term "radio" for both the receiving device and, more broadly, the wireless system of entertainment and information that emerged. "Broadcasting," moreover, was so firmly entrenched that it has never, except perhaps in agricultural circles, returned to the meaning of sowing seeds. Among the other unknowns in early broadcasting were factors such as the reach of the signals, audience reception and audience responses. For this reason, broadcasters would invite listeners to write to them detailing where the broadcasts were being received, on what frequency, at what time, and indicating the quality of the reception.

World War I provided some impetus for the development of broadcasting in the United States, though perhaps less directly so than in Europe. Austin (1922, p. 16) noted that the need for air-to-air and air-to-ground communications prompted the military to enlist the use of emerging radio technology:

> Came the war, with the urgent need for some form of rapid and positive communication between airplanes and between airplanes and the ground. The best radio talent was applied to the task, and soon little outfits, not much larger than one-foot square, appeared in our airplanes for the purpose of ensuring telephonic communication over some fifteen to twenty-five miles.

Not long after, military interests in the United States began to experiment with broadcasting aimed at amateurs and hobbyists involved in radio. The Anacostia Naval Air Station near Washington experimented with transmitting music on January 17, 1920:

> The Navy had, of course, been broadcasting by code for some years previously, but this was the occasion of the first radio telephone entertainment. . . . A phonograph concert was transmitted weekly . . . communications were received in February and again in March, 1920, from St. Louis, Minneapolis, and places in Pennsylvania congratulating the Anacostia station on the strength of its signals. (Austin, 1922, p. 16)

Schubert (1928, p. 192) suggested that, at least in part, the experience of World War I had expanded American perspectives to include the broader world and fueled the appetite for information, writing that "the use of every form of intercommunication had spread . . . the demand for news had stimulated the news-distributing agencies, both film and press, to unprecedented activity."

Frank Conrad and KDKA

Civilians and businesses were also involved in broadcasting experiments. Frank Conrad, an engineer at Westinghouse, made perhaps the most notable early impact. Conrad was not the only civilian involved in experimentation with radio broadcasting, just perhaps the most successful and well documented. Other efforts included such experiments as the October 27, 1920, "wireless telephone concert" played at the San Francisco Theatre for the listening pleasure of children warded at the San Francisco Children's Hospital using the Lee de Forest Wireless Telephone system ("San Francisco," 1920). The de Forrest apparatus with an antenna mounted atop the Humboldt Bank building (about a half mile from the location of the theatre in 1920) was in operation for several months and the so-called "musical messages" had been received as far away as Minneapolis, St. Paul and up to 1,200 miles at sea ("San Francisco," 1920). Crisell (1997, p. 14) marked the start of radio broadcasting (or the first demonstration of its power) with Dame Nellie Melba's singing performance from Marconi's facility near Chelmsford in the UK on June 15, 1920, which was said to be heard "throughout Europe and in parts of North America."

Despite evidence of multiple broadcasting efforts pre-dating and concurrent with Conrad's broadcasts, none were to gain as much of a following, command as much attention or demonstrate global reach as the Pittsburgh experiments in radio. Schubert (1928, p. 197) described Conrad's contribution and the following that he developed while working on a "special radio telephone transmitter for the Navy":

> He had set up an experimental station in his home . . . and as he made improvements and discoveries in his work, he tested them there by playing Victrola records. There was a radio audience, at first, of fifteen or twenty amateurs with home-made sets. They heard him once, twice—they began to listen for the Conrad tests.

These tests, according to Schubert (1928, p. 198), took on a life of their own and developed into something more than technical testing for Westinghouse on behalf of the US government as Conrad "found tests taken irresistibly out of his hands." He began to receive letters from listeners with criticisms, comments, encouragement, questions and even donations of records. His Wednesday and Saturday night broadcasts drew attention among listeners and the public. As Schubert (1928, p. 199) described, with the involvement of his employers, Conrad's broadcasts underwent an evolution from experiment to enterprise, particularly after news of the broadcasts got into the press:

> By the late summer of 1920 the Pittsburgh Department Stores had begun to advertise 'approved radio receiving sets for listening to Dr. Conrad's concerts.' And at that the Vice President of the Westinghouse Company, Harry P. Davis, who had been watching the playing of the whole little comedy, pricked up his ears. . . . A conception took shape in his mind. Why not make this more than Victrola records? Why not make it a "Newspaper of the Air," or something like that.

Before long, Conrad and Davis discovered a place for agricultural reports on the radio. Radio telegraph stations started sending US Department of Agriculture reports in December of 1920 but eventually, in 1922, switched over to audio radio broadcast with the distinct advantage that specialist operators were not needed to convert messages to and from Morse code (The Radio Staff of the Detroit News, 1922). As the *Wall Street*

Journal reported ("Market Reports to Farmers," 1921, p. 3), the "radio market news service of the Bureau of Markets, Department of Agriculture" had "arranged to broadcast market reports from Pittsburgh by radiophone."

By October of 1920, Conrad and Davis arranged to relay results of the November election. On November 2nd between five hundred and a thousand listeners followed the results from the *Pittsburgh Post* live on the radio station that would be known as KDKA. Wood (1992, p. 26) described the impact of the Harding-Cox presidential election coverage in cementing KDKA's place in history, being "usually cited as the historic beginning of regular radio broadcasting in America" despite the fact that KQW in California "first transmitted a radio program in 1909, and was running a regular schedule by 1912" and the Detroit amateur station W8MK "began regular transmissions a couple of months before KDKA."

Yet, for Wood (1992), Conrad's efforts at KDKA merit a historical place for being the first to broadcast on a daily basis to general audiences and to be licensed by the Federal Communications Commission (FCC) and for its important achievement of broadcasting the 1920 presidential election. Schubert (1928, pp. 199–200) credited Harry P. Davis with the idea of covering the election by radio in order to "launch the innovation as spectacularly as possible." Within a year of those efforts, the radio station would evolve from being a conduit for election results into also being a medium for campaign speeches and debates. An issue of the magazine *Electrical Review* of October 1921 recounted the station's role in connecting candidates with voters during Pittsburgh's mayoral election:

> Arrangements were made in Pittsburgh by the Westinghouse Electric & Manufacturing Co. to broadcast by radio the speeches by the candidates . . . thousands of persons were addressed at one time without the inconvenience of leaving their own radio set. Each candidate for mayor was sent to the broadcasting station, where he was allowed 5 min. to tell the reasons why he should be elected to the office. ("Electioneering by Radio," p. 673)

Austin (1922, p. 17) recounted that Westinghouse's KDKA station aired their first "radio-phone concert" on November 5, 1922, and later featured phonograph music and announcements and, eventually, live broadcasts. Sports also quickly became an important element of radio broadcast entertainment. Lescarboura (1922, pp. 60–61) described the introduction of boxing at KDKA:

> Casting about for features that would enliven the evening programs, it was decided to broadcast, as an experiment, blow-by-blow returns of a boxing match held in Pittsburgh. A private wire was installed from a boxing club to the radio station, and a man prominent in sporting circles engaged to render a round-by-round version of the progress of the fight.

Outside of the KDKA developments, interest in broadcast radio was becoming increasingly widespread as various radio clubs experimented with broadcasting and the popular press carried news of their efforts including a distinct focus on the distance of their coverage area and their ability to reach foreign audiences. *Billboard* (1921, p. 7) of May 7, 1921, for example, reported on the growing popularity and reach of experimental transmissions from the Union College Radio Club, a college group of radio enthusiasts whose weekly concert broadcasts were "heard by operators in 22 states" as well as by listeners "in Canada and ships at sea." This club later started their own station, WRL, in Schenectady, New York.

The same publication ("Music by Wireless," 1921, p. 32) carried news of the global reach of an American broadcast a few days earlier as a two-hour program of "imported" European dance music, duly Americanized, was broadcast such that it "carried to the ears of people in France and other countries." The broadcast was achieved through the facilities of a group called the Ship Owners Radio Service that started its own regular AM station (WDT-AM) in December of that year.

By 1922, in addition to a growing range of content, listeners (sometimes called "auditors") also had a growing range of stations to which they could tune. Lescarboura mentioned Westinghouse setting up a transmitter known as WJZ on the roof of its building in Newark, New Jersey, with studios in the lower floors from which they broadcast concerts and talks that listeners received in "Canada, Wisconsin, Florida, Cuba, and 600 miles out at sea." Westinghouse had established, by 1922, at least two other stations, a Springfield station, known as WBZ, aimed at New England and KWY in Chicago, aimed at midwest and western states.

Elsewhere, notably in Europe, radio broadcasting operations were evolving as well. Within a couple of years of the spread of radio in the United States, broadcasting stations began appearing on European airwaves. Periodicals in and around 1924 listed among these stations, broadcasters with regular schedules that could be received from numerous countries including The Netherlands, Spain, France, Great Britain, Switzerland, Germany and Belgium. The broadcasts included time signals, news, meteorological forecasts, agricultural prices in major markets (including, for example, livestock and fish prices at the Paris markets) and radio concerts.

Rixon (2015, p. 26) described reactions to early sound transmission efforts in Britain as focusing "more on its experimental nature, its non-broadcast uses, technical discussions and the sheer aural 'spectacle' or at least the wonder of hearing something live through a small box many miles from where it was happening," noting that the technology was "new, different and mysterious." Briggs (1961), Crisell (1997) and Scannell and Cardiff (1991) have all explored the uncertainty with which early

British radio approached the question of what was to be appropriate fare for British airwaves. Yet this uncertainty about the proper form of radio did not diminish enthusiasm for the practice and, particularly, for the allure of distance. Rixon (2015, p. 26) noted that British newspapers covered, with some interest in 1921, the fact that British Princess Mary listened to music transmitted wirelessly from a distant location, engaging in what was considered "the wonder of listening."

In the United States, as radio stations became popular, there was growing public interest in receiving this free entertainment over the air. A Radio Corporation of America (RCA) publication of June 1922 (p. 5) described radio as answering "the call for more liberal education of nations and peoples," while it permeated "the remote places of the earth," and claimed that the new technology was "sweeping the country" and "binding the people together in a new and democratic brotherhood."

But early radio's global reach and globalizing potentials were not limited to its signals crossing borders; sometimes rather than reaching for the globe, radio brought the globe closer to home. For example, evidence of broadcasting among ethnic and linguistic minorities early in the history of radio could be found in places such as the Spanish-language press of New York City. A 1922 article in a publication called *La Nueva Democracia* entitled *"Las Maravillas Del Radio"* (The Wonders of Radio) (James, 1922, p. 27), for example, referred to "El 'broad-casting'" as a system of transmitting messages through the ether and called it the most important advancement in radio, noting as well the prevalence of news, music, instructional programs (including agricultural extension courses) and Sunday sermons on the medium. By this time, radio's ability to reach remote communities was already worthy of mention as *"Las Maravillas Del Radio"* specifically mentioned Syracuse University radio courses, noting that persons living far from the cities could receive, through these programs, information surrounding the most modern methods of forest cultivation and conservation.

As radio broadcasting gained widespread notice and public interest, evidence of increasing signal reach also emerged. The *Hardware Dealers' Magazine* noted in January of 1922 that while Westinghouse was announcing a system of repeaters and stations to cover the entire United States after little more than a year of operation, its station at Pittsburgh was already enjoying wide and far-flung listenership, having "opened possibilities hitherto un-dreamed" by reaching "Canada, New England, Florida, Arizona, the Dakotas" as well as Cuba, Mexico and "ships in the middle Atlantic and on the Gulf of Mexico" ("Broadcast Radio," p. 136).

Similarly, a 1923 newspaper report described the range of General Electric's Schenectady station (in a period of still relatively little interference) as having been "heard in Cuba, Porto Rico [sic], Hawaii, Peru, Southern Canada, the Argentine and many other places far remote from their point of origin" ("Drama by Radio," 1923, p. 14). In 1924, the notion of "places far remote" took on even greater meaning for General Electric's Schenectady station WGY with a reception report from Cape Town, South Africa ("Radio Heard," 1924), a distance of some 7,880 miles. Notably, Cape Town would wait at least until January of 1925 to have access to signals from a formal station within South Africa. Even then, however, as the Western Electric station "JB" at Johannesburg launched with the sounds of a Zulu war dance, its signals were notoriously poor (Jollife, 1925).

Not to be outdone, WEAF, New York, boasted a reception report of their New Year's 1924 broadcast from an operator who listened in from the well-equipped Slangkop wireless station in Cape Town ("Radio Notes and Gossip," 1924), while, from the West Coast of the United States, General Electric's station KGO at Oakland, California, was boasting reception in the South Pacific and WGN in Chicago reported, in April 1924, being heard in Hobart, Tasmania, some 9,670 miles away ("Chicago Station," 1924).

Radio enthusiasts were always eager to demonstrate and report distant receptions, particularly since this ability to capture distant voices and entertainment was still relatively new and impressive. Whitefield (1922, p. 68), for example, described introducing these distant broadcasts to friends in the (then British colony of) Bahamas who, at first, considered the possibilities of such reception to be "a quaint piece of imagination" until transmissions from Miami, Newark and Pittsburgh were tuned.

Within a couple of years KDKA could boast not only local and regional success but also even greater international reach and acclaim. As Wood (1992, p. 14) put it, "KDKA's fame travelled well beyond America" as "KDKA became the radio station known to all radio enthusiasts in Europe." This increased reach was also a result of Conrad's use of short-wave repeaters and local retransmissions as *Scientific American* noted in May 1924:

> The feat whereby American broadcasting programs are repeated on these short waves and received and rebroadcasted by English stations, thus reaching the peoples of Great Britain, France, Germany, Belgium and the Scandinavian countries, is the outcome of two years' experimenting and perfecting of high-frequency apparatus. . . . Last October the Westinghouse Company inaugurated the first radio repeating station, known as KFKX, at Hastings, Nebr. ("Radio Notes: Short-wave," p. 357)

This was not a one-way flow, at least not for long. During the so-called "Radio Week" of November 1924, concerted efforts at international broadcasting, sponsored in part by *Radio Broadcast* magazine, caused major excitement among listeners on both sides of the Atlantic. The *New York Times* of November 26, 1924, reported that:

Broadcasting stations at Newcastle, England, Aberdeen, Scotland, and Madrid, Spain, were heard so clearly in America between 11 and 12 o'clock last night that devotees of radio reception had one of the most exciting hours of their lives. Local listeners-in by the dozens telephoned to The New York Times telling of their success in catching the overseas messages. ("Hundreds in City," pp. 1–2)

As experiments in broadcast radio evolved, so too did its range of content. Following the gramophone records and the election results relay exercise, KDKA embarked on live transmissions of church services and a range of other content, eventually offering "acts from theatres, musical recitals, reports of boxing contests, results of baseball, football and basketball games, complete minstrel shows, government market reports, New York stock market reviews," "national and international news" and "speeches of prominent men" ("Broadcast Radio Telephoning: Westinghouse Covering All the Country," 1922, p. 136). Among these "speeches of prominent men" on the radio was one on July 3, 1923, that caught the attention of the newspapers.

Dr. Hubert H. Harrison, an African-American intellectual was a "prolific writer, orator, critic, educator, and political activist" (Gaskins, 2009, p. 57) of the time. That he would deliver a speech was of little consequence as he was known to do so at exclusive gatherings of New York intellectuals as well as on street corners. However, newspapers took note that the radio broadcast of Dr. Hubert H. Harrison's speech on "the proper relations that should exist between 'The Negro and The Nation'" on behalf of the New York Board of Education, "through WEAF, the most powerful radio broadcasting station in the East," provided "an audience larger than any which was ever addressed by any black man before" ("Business Men," 1923, p. 6). Reports estimated the audience for WEAF at the time at more than 200,000 including listeners across the Atlantic in England ("Business Men," 1923, p. 6).

Other emerging stations were also known at the time to offer bedtime stories, agricultural information for farmers and concerts. Observers noted that the initial radio efforts at KDKA had not faded from public attention but instead garnered more interest over time. This endurance beyond novelty owed as much to an evolving range of content as it owed to expanding transmission reach as other stations came into being ("Westinghouse," 1921, p. 887). Among the emerging forms was one that would become a mainstay of early radio: the dramatic presentation or radio play. Anecdotal evidence of the impact of radio dramas spread internationally as a 1923 newspaper report would attest:

A woman's screams suddenly issuing from a dwelling recently caused an American policeman to demand entrance to the house to find out the cause of the disturbance. Much to his chagrin, he discovered that the offender was only a loudspeaker connected with a radiophone re-

cording of a scene from Eugene Walter's drama, "The Wolf," which
was being broadcast from the studio of the General Electric Company
at Schenectady, New York, many miles away. ("Drama by Radio,"
1923, p. 14)

In 1924, when the Eiffel Tower station aired a radio drama (adapted
from a Greek play), French newspapers were quick to congratulate them
on the achievement and to label it a first for radio. American newspapers
complained that the French probably had not yet heard of American
radio dramas ("Here and There," 1924). The British were also pursuing
the development of the radio drama. *Current Opinion* magazine described
the BBC's (British Broadcasting Company at the time) efforts in 1924 to
involve the audience in developing dramatic content with the offer of a
£50 prize for the best new and original play submitted for dramatic adap-
tation to radio:

> Since the actors cannot be seen by the audience it is important that the
> characters should be so well differentiated as to be easily distin-
> guished. The ideal length was set for this play at 20 minutes and it
> could not exceed half an hour. Not more than six characters were al-
> lowed. ("Travel and Communication," 1924, p. 380)

Later that year, the BBC appointed a new director to head up its dramatic
department and the company was said to be experimenting with background noises, the right voices for radio drama and writing that featured
only a small number of actors so that audiences could keep track of the
different voices (Cockaigne, 1924).

Several schools and colleges debated how radio might best be used to
supplement their educational offerings, and in several countries, radio
became a tool for everything from agricultural extension to adult literacy.
In the United States, the early development of radio did include some
instructional and educational content. Typical of these efforts was an
announcement from station WDAF in Kansas City, Missouri, in June 1924
that they would offer piano lessons by radio aimed at children with the
option to send in written work for grading ("WDAF to Broadcast," 1924).
The *Amaroc News* ("Plans Education by Radio," 1922, p. 1) carried a report from New York announcing that college education could soon "be
acquired at home through the wireless telephone medium" since New
York University announced "plans for the establishment of a broadcasting station at its Washington Square division from which classes in all its
courses" would be conducted.

Despite the evolution of these novel forms and the wide acceptance of
staples such as music programs, station managers and teams in charge of
programming admitted to being unsure as to what audiences preferred,
especially as programming began to reach further than local markets.
Broadcasting magazines surveyed station managers and programming
staff, while programming decision makers invited listeners to write in

expressing their preferences and even held competitions to encourage listeners to say what kinds of programming they preferred and to clarify their choices among options such as symphonic music or jazz.

Those becoming involved in broadcasting in the United States were a diverse group. Sarnoff (1939, p. 8) noted that "in those early days of broadcasting various organizations entered the field for the incidental advertising that could be obtained—flour mills, department stores, music shops, and even garages"—pointing specifically to the fact that "newspapers, too, were early entrants, foreseeing new possibilities for speedier communication." Among the newspapers entering the radio broadcasting arena by 1923 were "the Detroit News, the Detroit Free Press, the Kansas City Star, the Atlanta Journal, the St. Louis Post Dispatch, the Rochester Democrat and Chronicle, the Louisville Courier Journal, the Fort Worth Star Telegram, the Chicago Daily News, and the Los Angeles Times Mirror, among others" (Hilmes, 1997, p. 51).

When the *Detroit News* newspaper opened its experimental audio news transmission service in 1920, the broadcasters were still wrestling with terminologies and referred to their own efforts with labels that included "wireless telephony," "radiophone" and "radio." In a 1922 publication (The Radio Staff of the Detroit News, 1922, p. 7), the paper called itself "the first newspaper in the world to install a radio broadcasting station, and the first to increase its social usefulness by furnishing such a service to the public." Similarly, on March 29, 1924, the *Chicago Daily Tribune* newspaper announced that it was launching its radio service W-G-N Tribune radio through acquisition of a local broadcast station, WJAZ, at the Edgewater Beach Hotel promising news, music and other entertainment ("W-G-N, Tribune Radio," 1924). The station's launch publicity boasted the inclusion of five hours of what it called an experimental broadcast entitled *Across the Pacific* involving an arrangement with station 4YA in New Zealand ("W-G-N, Tribune Radio," 1924, p. 1).

The *Ventura Free Press* (1932, p. 41) described the public reaction to this new form of entertainment in which, early on, the thrill of receiving a signal outweighed any qualms about content:

> In 1921 "listening in" became America's chief indoor sport. The public seized upon radio with even greater ardor than it had welcomed the bicycle, horseless carriages, ping pong, the movies, auction bridge. Radio eclipsed all previous fads. More and more stations sprang up about the country. More and more receiving sets were installed. . . . The pioneer radio fan, listening through the ear-phones of the clumsy apparatus of the day, was not fastidious about his programs. What he heard was of secondary importance. He got his thrill from fishing things out of the air, and the greater the distance spanned, the greater the thrill.

While early broadcast radio was developing in the United States, around the world, other communities were also exploring the potentials

of the new medium. In China, during the 1920s, newspapers were starting to acknowledge the promise of radio as a tool of local information
and as a global window. Radio technology was becoming cheaper and
more accessible in 1920s, rural and amateur enthusiasts had already tak-
en to the airwaves in China. Shangai was one of the main centers of early
broadcasting in China with E. G. Osborn (an Anglo-American entrepren-
eur and journalist) setting up a radio station in 1922 with its inaugural
broadcast in January 1923. Various estimates place his audience at
around five hundred sets at the time—many sold in anticipation of the
broadcasts. Osborn was part of the Shanghai Amateur Radio Association
formed in February 1924 comprising "both foreign and Chinese members" ("Shanghai Radio Assocn," 1923, p. 25) and convened at the offices
of the *Shanghai Times*.

The *China Press* newspaper partnered with the American Kellogg
Switchboard Company (Krysko, 2011) to establish the China
Press–Kellogg Radiocasting service from Shanghai that conducted its first
broadcasts in December 1924. In January of 1925, the newspaper touted
the impact of its broadcasting enterprise, noting that they were receiving
letters and telegrams indicating reception as far away as Hong Kong,
Tientsin, Hangchow and even boats off the coast of Yokohama, adding
that:

> Now, scarcely a day passes but that *The China Press* receives letters
> from owners of radio sets in places such as Soochow, Nanking and
> elsewhere in the Yangtze Valley expressing their appreciation of the
> pioneering work which this newspaper, in cooperation with the Kel-
> logg Company, is doing in this important field of endeavor. There are
> reported to be some three or four thousand radio receiving sets within
> reach of the station here and more are being installed rapidly. ("Radio-
> Broadcasting in China," 1925, p. 10)

The newspaper described its radio station offerings (received, it claimed,
"by sets all over the Orient, from Harbin to Canton, and from beyond
Yokohama to the Philippines") in terms of news of current happenings in
China and abroad as well as lectures and musical programs, emphasizing
the global possibilities such that "interior places inaccessible to the rail-
road or even the telephone and telegraph" were "brought into direct
communication with the outside world" ("Radio-Broadcasting in China,"
1925, p. 10). These broadcasts included both English and Chinese programming with market reports, stock reports, speeches, music and news
("Listen In," 1925, p. 2).

In February 1925, the Chinese government announced it would establish radio broadcasting facilities at Peking, though Shanghai was still to
be a focal point for radio; information was to be sent from Shanghai to
Peking via telephone and wireless. The Chinese government broadcasts
were to primarily be in Chinese with about a third of the content being

English ("Government at Peking," 1925). Radio Peking, founded in 1950, would eventually become Radio Beijing, China's international radio broadcaster—eventually known as China Radio International (CRI).

Early in the development of broadcast radio, China was also seeking distant signals including those of American stations such as WGY at Schenectady. Yet, despite all this attention to radio broadcasting, not everyone was enthusiastic about the "local craze for radio-telephony" in China (Humana, 1923, p. 528). A letter to the editor of the *North China Daily News* in 1923 lampooned the phenomenon, suggesting that Chinese radio programming should be expanded to include such unlikely content as "passengers expectorating in the first class compartments of the trams," flute solos "played entirely with the left hand" and the sounds of chewing at dinner (Humana, 1923, p. 528).

Alongside the development of global perspectives in radio broadcasting in the United States and the rest of the world, much darker content was also emerging. On a broad scale, propaganda from the Fascists and the Nazis in Italy and Germany would receive much attention in the years to follow. As Boyd (1999b, p. 101) pointed out, widespread belief in the power of the relatively new medium and the evidence of radio advertising motivated the use of radio for persuasive political broadcasting:

> Broadcasts coming from the ether were thought to be a very effective means of communication, especially if the opposition controlled or influenced state-run or -sanctioned radio and newspapers. It was Lenin who reportedly observed that radio was a "newspaper without paper and without frontiers."

Russian broadcasting was indeed rooted in authoritarian philosophies and, as Lovell (2013, p. 80) has ably outlined, used various methods including public loudspeakers and wired systems along with over-the-air radio to support government discourse from the early 1920s, offering "a way of projecting the voice of authority into every workplace and communal flat in the USSR." Lovell (2013, p. 80) noted in particular that "in the 1930s radio spread across urban Russia principally by means of 'wired' networks that allowed the audience no discretion to switch channels (or even, in many cases, to switch off)."

TWO

US Radio, the Global and the Local

Folami (2010, p. 141) argued that "at radio's mass emergence, many perceived it as the vehicle through which America's locally, regionally, ethnically, and/or socioeconomically marginalized populations could be included in America's democracy by being given an expressive and deliberative space on this newly accessible and fairly inexpensive medium." Indeed, as radio first emerged, transmitters reached only local areas and transmitted material of substantially local interest (such as sermons from local churches and musical performances from local concert halls). Yet, the national and global dimensions were never far off. Stations were quickly able to transmit their signals to much greater distances and the allure of receiving national or international news via a local station soon expanded the scope of the material that could be offered on a day's broadcast. Krysko (2007, p. 335) explained that these broadcasts would include foreign language programming within the United States, often on stations owned and operated "by an array of ethnic, religious, educational, and labor groups."

Various reports estimated foreign language broadcasts at some 1,500 hours weekly on 100–200 stations nationwide during the early 1940s (Barnouw, 1968; US House of Representatives, 1944). These included broadcasts of news in foreign languages on major mainstream and network stations. As *Radio Daily* reported, an existing news provider, Van Cronkhite Associates Inc., announced a service in 1937 that would provide foreign-language news in Italian, German, Polish and Portuguese "delivered direct by teletype printers" including headlines, "special feature material, sports, women's items, oddities, etc." ("Language News Programs," 1937, pp. 1, 3).

Despite their popularity, these stations (or occasional broadcasts) were subject to both suspicion and derision, accused variously of ham-

31

pering immigrant assimilation (Arnheim & Collins-Bayne, 1941), polluting the airways, fostering foreign values, and of collaboration with enemy forces (Barnouw, 1968; Krysko, 2007). Fay (1999, p. 63) described fears at the outbreak of World War II that "German and Italian announcers, under the influence of their mother countries, could sway immigrant listeners toward pro-Fascist propaganda" and could "leak military information to the Axis countries through radio transmissions that could be picked up abroad."

Such fears eventually led to a decline in foreign-language broadcasts on US radio, particularly after the onset of World War II, despite broadcasters' attempts to monitor content and avoid even the appearance of foreign sympathies. The ominous presence, as well, of the government's Foreign Language Wartime Control Committee, which introduced a "voluntary code of wartime practices" (including an undertaking to examine the backgrounds of anyone involved in producing foreign-language material) (Lang & Simon, 1942, p. 77), did little to encourage stations to continue with these broadcasts. Horten (2003, p. 70) described "popular suspicions of a fifth column working inside the United States" suggesting that "even before America's entry into the war, there was a marked decline in foreign-language broadcasts."

In addition to domestic foreign-language radio programming, there remained, particularly in the early days of US radio, a fascination with receiving broadcasts from distant places. Krysko (2007, p. 335) noted that "while foreign audiences enjoyed radio entertainment in their countries, these same programs also reached the American audiences listening at precisely the same time through the sheer force of the transmission signal."

Graves (1941, p. 11) argued that the spread of international broadcasts within the contexts of war and peace had become "as much a mark of nationhood as the maintenance of armies and the right to conclude treaties." Even prior to the United States' involvement in World War II, noted Graves (1941, p. 8), several powerful players in global geopolitics were "wooing the United States" with their broadcasts:

> The US early in 1941 was served with three and a half hours of broadcasts a day by the Rome radio, with six and a half by the BBC, and with nearly eleven by the German radio. As crisis mounted in the Far East, Japan increased her North American service to four and a half hours a day.

These deliberate services provided only a portion of the available broadcasts. According to Graves (1941, p. 8), at that time listeners were able to hear many incidental broadcasts which were "simply sent to the United States at the same time as they are transmitted to the audiences for which they were composed" such as German newscasts to England that were sent simultaneously to the United States. During a given day in the pre–World War II continental United States, related Graves (1941), an average listener in the United States might receive news from London in English (both directed at the United States and also incidentally from programming to the British Empire) or in French (intended for France or Belgium) or in German (intended for Germany) as well in several other languages.

They might also receive programming from German stations in Spanish broadcasting to Spain, in Afrikaans to South Africa and in English to England; Japanese broadcasts to places such as Australia could be received in English; Italy could be heard broadcasting to its neighbors in Europe as well as to Arabia and Egypt and a Spanish-language service as well as Portuguese and Italian language broadcasts bound for Latin America. These broadcasts did not only emanate from major centers of world power. Radio Algiers in French-controlled Algeria was at the time broadcasting a signal primarily intended for France, but which could be received as far afield as the United States. In addition to their geopolitical targets for propaganda and suasion, these international broadcasts also included transmissions deliberately aimed at ethnic communities. An example of this can be found in the fact that German radio stations were known to broadcast programs of German folk music intended for diasporic and ancestral German communities in the United States (Graves, 1941).

As radio became more commercial in the United States with the introduction of advertising and the role of large powerful corporations such as AT&T and the Radio Corporation of America, attention to foreign broadcasts and local broadcasts to marginalized populations became less of a priority. Stronger, more reliable local signals were much more listener-friendly and provided much of the information and entertainment that would previously be obtained from squelching, fading and otherwise fickle distant sources. Content of broader appeal more clearly suited stations intended for broad audiences and wide advertising reach. Where rural or local markets were considered, it was primarily with a view to providing such areas with national broadcasts through the power of the networks.

Corporate pioneer of radio (and later, television) at RCA and NBC, David Sarnoff, noted (1939, p. 8) the dynamic tensions between the global and local potentials of radio broadcasting from its earliest manifestations in the United State as RCA strove "to provide an American-owned system of international communications" in which "the devices and patents which made it possible for RCA to operate an international communications system also were required to make home receiving sets and broadcast transmitters."

CHAIN BROADCASTING AND NETWORKS

Challenges of distance and geography played an important role in the form of the early radio industry in the United States. The *Electrical Review* ("Westinghouse," 1921, p. 887) described efforts from industrial giant Westinghouse to cover the country with radio programming that would allow "anyone anywhere in the country to enjoy the many benefits of radio." Westinghouse touted the fact that KDKA had been able to reach not only distant parts of the US continental landmass but also locations abroad, including Cuba and Mexico ("Westinghouse," 1921, p. 887). This early vision of covering the entire United States with radio involved the development of other Westinghouse stations to serve different regions that KDKA did not reach with sufficient strength including stations at Springfield, Massachusetts, and at Newark and Chicago.

Once listening moved beyond the thrill of distant signals and audiences evolved into consuming content, radio stations in the United States were faced with the challenges of content production and reliable signal propagation to their audiences. As Sarnoff (1939, p. 8) noted, "it soon became evident that the growth and permanence of radio broadcasting depended primarily on the quality and variety of programs" as "the novelty of tuning in distant call letters quickly wore off."

The early American listener might be faced with limited signal options; high-quality broadcasts might only be available from distant stations whose signals would be weakly received. The solution to this problem in the vastness of the US landmass was something called "chain broadcasting" that evolved under what came to be known as radio networks in the United States. The Federal Communications Commission's *Report on Chain Broadcasting* of 1941 (Federal Communications Commission, 1941) described the system that accounted for over $46 million of business (among NBC, CBS and Mutual) as "simultaneous broadcasting of an identical program by two or more connected stations" usually achieved through telephone line links. The FCC (1941, p. 4) noted that "the growth and development of chain broadcasting found its impetus in the desire to give widespread coverage to programs which otherwise would not be heard beyond the reception area of a single station" as well as making possible "a wider reception for expensive entertainment and cultural programs and also for programs of national or regional significance."

Sarnoff (1939, p. 9) described the idea of connecting stations through existing telecommunications infrastructure and development of the network system as "a democratic answer" to government control that was "found by private enterprise" when, in 1926, RCA took over an experimental system of interconnecting radio stations from the American Telephone & Telegraph Company (AT&T), extending it to a group of independent stations.

Radio networks evolved to allow content sharing and provided a method of covering the entire nation or significant parts of it. As Cox (2009) has noted, audiences in the 1940s and 1950s much preferred to listen to network broadcasts than those of independent stations. Networks were able to pay for better content and provided more of a national or regional perspective. For many decades, this remained the case, and the powerful radio networks, later to be joined with television operations, provided both national and international perspectives to US audiences.

Radio networks were able to connect advertisers, content producers and audiences instantaneously across space, creating an economic model that would prove successful in the years before and after World War II. Network radio flourished in the United States driven by commercial forces as a free market enterprise and with minimal government oversight as to licensing and frequency allocation. The Chain Broadcasting Order emanating from the Report on Chain Broadcasting (Federal Communications Commission, 1941) prohibited networks from owning multiple stations in single markets and from otherwise dominating local markets. Despite these limitations, however, the major radio networks still allowed advertisers to reach a national audience, including even rural or remote areas, through their affiliates.

With reliable, strong signals and high-quality content available on the radio from local stations, there was little need for audience members to invest time or effort into distant listening. Several scholars have argued that these network broadcasts were responsible for not just creating a mass consumer audience base, but also, in diverse ways, contributing to national cohesion and cultural homogenization (Bagdikian, 2004; Cohen, 1990; Hilmes, 1997) in the sense that might later be described as national identity (Deutsch, 1953) or in terms of imagined communities (Anderson, 1983). Isaksen (2012, p. 755) has argued that while "the purpose of radio, from its inception, has always been to entertain, inform, and turn a profit" the medium was also "influential in building a sense of nationhood by bridging the physical distance between its widespread listeners."

Smulyan (1994, p. 31) described the evolution of national broadcasting as a consequence of prevailing economic, social and technological trends in the 1920s since the large companies involved in radio "had a national outlook and an interest in broadcasting to the entire country" and due to the notion that "intellectuals and ordinary Americans shared a belief that new communications technologies should draw the United States together." This very tendency toward homogenous national programming under powerful national corporate interests also had its critics who, during the rise and ascendency of the radio networks, decried the networks' monopolistic tendencies and their almost inevitable overshadowing of local and parochial news and entertainment in favor of homogenized national fare. Folami described the situation as one in which radio programming was a "slave to corporate control" (2010, p. 144).

US RADIO AND WORLD WAR II

During World War II, US radio networks brought news of the world and a global perspective to audiences in the United States. The evolution of the so-called "war correspondents" came from the deployment of foreign correspondents on behalf of the networks as tensions built toward war in Europe. According to McCormick (1953, p. 24), these correspondents served, despite serious obstacles, as a primary conduit of information to the US radio audience:

> In 1939 the National Broadcasting Company—then two networks under one management—employed 46 reporters scattered around the world. The Columbia Broadcasting system had 24. Mutual had six. . . . Although what they said was tempered by censorship, they were heard by millions of Americans who would otherwise have received a minimum of foreign news.

"In mid-1940," wrote Barnouw (1968, p. 140), "the American networks were making some twenty foreign pickups a day." Once the United States became involved in World War II, the movement of information became much more controlled. On January 16, 1942, the Office of Censorship issued a Wartime Code of Practices for broadcasters urging voluntary actions to support the war effort. On that same day, President Roosevelt named the Office of Facts and Figures (which was replaced by the Office of War Information) as a "clearing house for Government broadcasting" ("Radio Highlights," 1943, p. 56). Soon after that designation, US radio broadcasters formed the Broadcast Victory Council to act as liaison with the government on behalf of industry.

The Office of War Information (OWI) utilized the services of the four radio networks to broadcast programs about the war effort. These efforts included four separate schemes intended to inform and persuade audiences through programming carried on the networks. McCormick (1953, p. 24) argued that "if the radio division of the OWI was a grandiose scheme for wholesale manipulation of public concern, it was too confused to be successful" since "the senses of the American people are not easily dulled into mass subservience," noting, as well, that "straight, factual, intelligent news continued to be broadcast night and day."

While the OWI ostensibly encouraged involvement of the networks and their leadership, claiming that this was preferable to direct funneling of information (Landry, 1943), others considered the OWI's methods to be anathema to free speech. When the OWI director, Elmer Davis, announced a series of 15-minute broadcasts to be carried on the national radio networks "to summarize and clarify war developments" in March 1943 ("OWI Director," 1943, p. 5), not everyone was convinced of its value to the war effort. In a public statement Senator Robert A. Taft (R-Ohio) condemned Davis' radio programs as "propaganda broadcasts" that amounted to "an abuse of power and an insult to the intelligence of the American people and to the abilities of our free press" ("OWI Radio Time," 1943, p. 3). Taft went even further, commenting that "Americans do not take kindly to propaganda by their government" and that "they have learned to despise the methods of Goebbels in telling the German people when to listen and what to think" ("OWI Radio Time," 1943, p. 3).

OWI officials insisted that they were not involved in commandeering airtime, citing the eagerness of radio networks and independents in assisting the war effort ("OWI Radio Time," 1943; Landry, 1943). A later director of OWI, George P. Ludlam, called the OWI's collaboration with American radio "a joint industry-government weapon of war" and estimated the contributions in time and talent from the industry at about $55 million in 1943 and about $66 million in 1944 (Ludlam, 1945, p. 18). Indeed, NBC International boasted about receiving letters from US soldiers around the world who were listening to "Bing Crosby's baritone and Bob Hawk's puns" in places including "Iceland, Canal Zone, Trinidad, British Guiana, Brazil and Hawaii" ("NBC Soldier Letters," 1942, p. 47).

In addition to the OWI's broadcasts and contributions from the networks, the global dimensions of the United States' wartime radio efforts were also evident in the attention paid to foreign-language broadcasting services. Barnouw (1968, p. 157) described the involvement of these foreign-language broadcasters as being riddled with difficulties with some carrying "broadcasts in as many as ten languages":

> No official at the station could understand half the programs it was broadcasting. The complexity of this problem was illustrated by the case of an Italian announcer at a foreign-language station who was said to echo Axis propaganda even while dutifully urging the purchase of United States war bonds.

The US government not only engaged in various efforts to broadcast its own messages on radio aimed at domestic and foreign audiences, it also engaged in broadcast monitoring activities. "Early in 1941," wrote Barnouw (1968, p. 157), "the FCC set up a division to listen to and analyze foreign short-wave broadcasts. After Pearl Harbor it acquired special importance and was named the 'Foreign Broadcast Intelligence Service.'"

After the end of World War II, the US government remained concerned about international broadcasting from the United States. Several commercial operations aimed at international audiences found themselves facing government pressure. The World Wide Broadcasting Corporation of Scituate, Massachusetts, for example, found itself in conflict with the FCC over its international broadcasts. The corporation had been involved in international broadcasts from as early as 1935 and the government seized its facilities during the war for military use. After WWII, the US government leased facilities such as the four stations run by World Wide and allowed the stations to reclaim a portion of the time for their own use. The FCC raised at least 12 objections to World Wide's operations over concerns dating back to wartime and their investigations continued for several years. The general thrust of government intervention into US international broadcasting was to discourage commercial efforts. Eventually, only government and religious organizations continued to pursue international broadcasting from the United States.

The global perspective that World War II necessarily prompted continued to impact programming on American commercial radio even after the war. In 1946, for example, CBS producer Norman Corwin undertook a journey around the globe to interview leaders and everyday folk alike on their perspectives surrounding the war and its aftermath (Barnouw, 1968; Keith, 2009). Interviewees included schoolchildren in various countries including Egypt and the Philippines as well as leaders such as Jawaharlal Nehru in India and the British minister of state, Philip Noel-Baker.

As the war ended, resources could be once again directed to the development of television and the networks were quick to bring television service to audiences in the United States. Television brought major changes to how audiences listened to radio and the era of network radio had to yield to television in the 1950s and 1960s. Sterling and Kittross (2002, p. 276) suggested that the return of soldiers from the war with radio training was one of the pressures brought to bear on US radio in the post-WWII years as these former soldiers sought to establish their own AM stations in the emerging markets.

By the early 1960s, commentators such as Stuart (1963, p. 4) began to note changes in radio, "the dominant mass entertainment medium of the prewar era," that included a rise to prominence of independent stations and a fading of the networks in the face of competition from television. Perhaps the best evidence of these changes was the movement of money from network radio to local radio (Stuart, 1963). In the face of the networks' television operations dominating the national and international news and entertainment, radio was forced to become more local. Sterling and Kittross (2002, p. 283) pointed out that at the end of WWII in 1945, 95 percent of all radio stations on the air were network affiliates, but by 1952, that figure had dropped to just over 50 percent as television became increasingly powerful so that "between these two developments, radio as a national advertising and programming medium gave way almost totally to television." Hilliard and Keith (2005, p. 56) also described this shift away from the network-driven national content that prevailed up to and just following World War II as "local stations became more and more involved in programming that originated locally."

The move to specialized music or talk stations rather than generalized network programming is often termed the format era in which a listener tuned in based on the choice of listening to country music or oldies or talk. For some commentators, such as Knoll (1968), this era marked the erosion of both radio journalism and the formal radio news program, which took a back seat to the ascendant television newscasts. Even entertainment programming such as variety shows and soap operas, initially hits on radio, migrated to the visual medium. Both national and global perspectives quickly became the domain of television, which could provide not just narrative accounts but also the verisimilitude of distant locations and events both domestic and foreign.

Radio stations, whether affiliates or independents, eventually were able to obtain high-quality programming outside of the bounds of network agreements through a system known as syndication. Under syndication, programming would be licensed to a radio station independent of any network agreements. Thus, an independent could obtain, for a fee, the right to broadcast a successful network program. While this diversified options in the United States, syndication also had global ramifications.

By 1994, some 88 percent of stations in major media markets were using syndicated programming within the United States (Borzillo, 1994), but these programs also found their way overseas. By the early 1990s, American programs were being beamed via satellite links to foreign markets by arrangement with US media interests, but this was nothing new. For many decades, US radio content (even discounting the immense popularity of American music) had been making its way into the programming schedules of radio stations around the world. Stations in far-off places such as Trinidad and Tobago were able to secure (for a fee) vinyl discs with programming such as ABC Radio Network's *American Top 40* during the 1970s. Earlier, soap opera classics such as *Guiding Light* featured on the broadcast schedule of many local broadcasters around the world. In some cases, US radio content was reproduced in foreign languages for syndication in different linguistic markets as well (Borzillo, 1994). Even in English-language markets, the influence of US radio soap operas could be felt, though sometimes indirectly. Goodman and Smulyan (2013, p. 164), for example, described a form of American radio export to the international market that they describe as somewhat hidden from historical view, namely "the transformation of American radio serial scripts—lightly re-written but completely re-voiced in Australia—into successful export commodities in the 1950s and 60s."

In this way, even though US commercial radio receded from a direct role as global broadcaster, it maintained a global presence through content production and syndication. So widespread was US radio industry influence that even the Soviet Union eventually yielded to its power. In 1990, American radio took root on Soviet soil as US radio interest Westwood One—which would eventually become one of the most powerful corporations in American radio broadcasting—began to infiltrate Soviet air on a commercial (and cultural) basis. Garber (1990, p. 1) recounted that:

In January 1990, Westwood One Chairman and Chief Executive Officer, Norman J. Pattiz, announced an unprecedented agreement with Gosteleradio to become the first American company to produce original, regularly-scheduled, commercial programming for broadcast on Soviet radio. The agreement called for Westwood One to create and produce three radio series, and in early February, Westwood One debuted USA Top 20, America on Record and American Musical Classics, a monthly series that explores the subject of American classical music.

Thom Ferro, Westwood One's executive vice president and general manager, described reaction to the shows from listeners in the USSR at the time as "overwhelming," even suggesting that their mailbags had been "overflowing with letters from listeners" thanking them for sharing the "American music and lifestyle with them" (Garber, 1990, p. 1).

DEREGULATION, CONSOLIDATION AND HOMOGENIZATION

Later restructuring of the media environment and legal changes (notably, relaxation of FCC ownership rules in 1992 and passage of the Telecommunications Act of 1996) led to a further devolution of the radio network model with corporate conglomerates buying up and operating stations throughout the United States. Folami (2010, pp. 142–43) noted that "deregulation of the media industry which began in the early 1980s and was solidified by the Telecommunications Act of 1996, facilitated unprecedented consolidation in radio station ownership" resulting in a radio industry that "has become a commodified and commercialized wasteland—a corporatized plaything—littered with fragmented yet overlapping music formats that play the same homogenized corporate-produced music playlists and are devoid of meaningful local public- and cultural-affairs programming." Lasar (2016, p. xii) described this as the "Clear Channel approach" citing "its emphasis on centralized, automated programming and the decline of the locally based independent radio station."

Not everyone agreed with this negative view. Even some industry voices expressed support for the emerging model as, in search of efficiencies of production, corporate interests bought up and consolidated local stations, combining facilities and staff, often operating multiple stations from the same building and using resources across these stations. In 1999, while acknowledging rapid ownership consolidation and the fact that the top seven owners in the United States accounted for 40 percent of all industry revenues, management publication *McKinsey Quarterly* praised the virtues of consolidation and noted the potentials of emerging technological developments (Alderton, Krim, Schmitt & Sheehy, 1999, pp. 124–25) "that make it possible to operate stations remotely—or even to automate them entirely for large portions of the day."

One of the early cost-cutting strategies in this wave of ownership consolidation was to combine news gathering across the former individual stations into single newsrooms or, in some cases, single individuals. Eventually, many stations eliminated their news operations and limited their news content (if any) to relays of news from external sources. During the early 1990s industry surveys revealed steep declines in radio news staff (often accompanied by closing newsrooms) and an increasing reliance on syndicated news services as super-specialized music and talk formats dominated radio (Adelson, 1994).

Torosyan and Munro (2010, p. 34) argued that both consolidation and what they term "outsourcing" ("importing news, weather, sports, traffic, and other content from distant places to save costs locally") have eroded the ability of radio to contribute to its local communities in the United States and "led to concerns about the loss of 'local identity.'" The increasing availability and diminishing costs of satellite linkages and broadband Internet connections have enabled modern corporate networks as well as local independents to engage in low-cost homogenized programming emanating from one central location and syndicated to many stations at the same time. Programs ranging from Saturday night dance programs to love and relationship advice with music air in cities ranging from Albuquerque, New Mexico, to Poughkeepsie, New York. Local radio stations in such town, to this day, receive mail for personalities like Delilah whom listeners perceive to be in their own local communities (though her advice and music program emanate from Seattle, Washington).

THREE

From Chance to Design, the Rise of International Radio Broadcasters

As radio broadcasting evolved across the globe, many broadcasters became inadvertently global in their scope, often surprised at reports of their signals being received in places much further away than they intended or imagined. This inadvertent global scope, accompanied by enthusiastic distant signal listening and even the formalized re-transmission of distant signals, eventually gave way to deliberate efforts at global broadcasting from stations and organizations dedicated to that purpose. The present chapter examines the evolution of global scope from chance technical overspill to commercial opportunity to tools of political and religious propaganda.

EARLY DISTANT LISTENING

During the 1920s and 1930s, radio-enthusiast listeners far away from the early centers of radio broadcasting such as Pittsburg's KDKA and the Dutch PCJJ station at Eindhoven tuned in to try to capture these elusive distant signals. With some investment in antennas and receiving equipment and with cooperation from the weather and other atmospheric factors, they might successfully receive broadcasts from afar but often faced problems including signals that faded in and out. Sound quality was often poor even if a signal was strong. Later, local authorities either established facilities or, in some cases, adapted existing technology to facilitate listening. In British Guiana, for example, local colonial authorities adapted existing antennae and signal amplifiers on the east coast near the capital of Georgetown previously used for wireless Morse code to serve as relay systems for distant audio broadcasting signals.

Radio listeners outside of the main centers of broadcasting were able to enjoy programming from distant stations in other countries and even other empires. Guyanese radio listeners in the 1930s were able to tune in regional stations such as the Barbados broadcasting station YP6YB but also listened to broadcasts of programs from the BBC such as James Laver's reading of his *Nightmare Series No. 7* and presentations with titles such as *A Concert Party Entertainment* that featured various band presentations, often live from locations in England. At the same time, they could also choose to tune in to programming from General Electric's stations W2XAD and W2XAF in Schenectady, New York, for programming such as *The Sizzlers* (a musical trio), the *Vic and Sade, Comedy Sketch,* show (popular in the United States from 1932 to 1946 and later adapted to television), soap opera programming such as *Oxydol's Own Ma Perkins, Dramatic Sketch,* featuring Virginia Payne as Ma Perkins and *Dorothy Page Songs* from the so-called "Singing Cowgirl" ("Radio," 1935, p. 8).

With the formalization of radio enterprises came awareness of the promises and potentials of broadcasting messages widely—including to those far removed geographically. Thus, long-distance or global broadcasting evolved from the simple and often accidental or fortuitous reception of far-away signals into purposeful transmission of signals to distant audiences for various purposes that included political influence and proselytization.

Harnessing Global Reach

Saerchinger (1938, p. 344) argued that the development of broadcasting, particularly in Europe, was tied to World War I and governments' attempts to harness, control and even limit the potentials of wireless in its various manifestations, writing that: "the governments of Europe, once their hands were on radio, first refused to loosen their grip, and then continued to hold over it a 'protecting' hand, which was later to tighten into a stranglehold." First in France, then in one country after another, the authorities opposed the introduction of broadcasting by radio. Severe restrictions were placed in the way of amateurs: playing with radio waves which travelled across frontiers as easily as within them was—to the war mentality—worse than playing with fire.

This and other differences in the evolution of the medium between Europe and the United States help to explain why the imperial powers of Europe pursued government-owned and -controlled media (often driven by the imperatives of serving far-flung colonial interests) while American radio developed primarily as a commercial enterprise with a primary focus on developing local markets but with distant reach as a desired outcome.

KDKA in Pittsburgh, with its shortwave repeater, for example, counted the residents of various European capitals among its audience members. Berg (2013, p. 29) described the initial evolution of the KDKA signal into a global phenomenon:

> Starting in the summer of 1922 Westinghouse had discussions with A.P.M. Fleming, manager of the research department at Metropolitan-Vickers Electrical Company of Manchester, England, who was then in the United States, about a plan for Metropolitan-Vickers to attempt to pick up the KDKA shortwave signals and relay them to its listeners. . . . Success was achieved in September 1923 when KDKA shortwave was received at the Metropolitan-Vickers facility at Altrincham, Cheshire.

Fejes (1986) argued that, with successes such as these, experimenters and inventors in early radio upended scientific orthodoxy that doubted their ability to reach distances on a global scale with their audio broadcasts. The *Chicago Daily Tribune* of November 9, 1924, took note of the (often unexpectedly distant) reach of early broadcasters:

> There's a man in the jungle, 200 miles from the nearest habitation in British Guiana, who, when he asks his supervisors off in Georgetown for a short vacation for a brief rush with civilization, is always told "you get more amusement out there than we do here in town." The reason is this: H. Johnstone Smith, Inspector of police at Morewanna, British Guiana, is constantly entertained by concerts broadcast from KDKA at Pittsburgh 3,000 miles away. ("Hears KDKA," p. F10)

Europe placed a similar focus on achieving transmissions over great distances and evaluated radio broadcasts in terms of experiments to bridge such obstacles as the Atlantic Ocean (Saerchinger, 1938). Since broadcasters and radio enthusiasts of the time valued the ability to reach listeners as far away as possible, broadcasters solicited reception reports, boasted of the distant reaches of their transmissions and adjusted their equipment to maximize range. Similarly, audience members wore as a badge of honor the ability to receive distant broadcasts—often a function of investment in more powerful receivers and larger antennas, but sometimes also a function of atmospheric conditions and pure luck.

Fortner (2005, p. 3) privileged listening audiences' excitement about receiving distant content over any concerns about the actual content itself, writing that for early listeners, "everything was a novelty—tastes in music or topics for interviews, lectures, or discussions were not relevant considerations." What mattered more was the ability to connect, through listening, "to the cultural products and political rhetoric of other countries without having it filtered by the local press or political establishment" (Fortner, 2005, p. 3). According to Sarnoff (1939, p. 8), "the public sat up late at night to capture the faint, elusive call letters of distant stations" and it was this "thrill of listening to far-off places that gave radio broadcasting its first impetus."

This thrill of listening to far-off places was no less evident in some of those "far-off places." Radio listeners in India in the 1930s, for example, were already receiving a mix of local, colonial and international content. Indian listeners could consult published listening schedules for the times and frequencies of broadcasts from major urban centers in their own territory as well as programming from further afield, including broadcasts from Nazi Germany ("Radio Broadcasts," 1936). The allure of distant reception was so great that some broadcasters suffered from the problem of fake reception reports sent by overzealous fans. A 1936 issue of *All-Wave Radio*, for example, recounted that the New Zealand Radio Association had released data on a rash of reception reports that could not possibly have been true since, in at least one case, the station reported had been off the air for several months (Hinds, 1936).

The global dimensions of radio and its ability to easily cross international borders also made for neighborhood squabbles. Charlesworth (1935) described concern among Canadians over their radio (in and around 1929) being overwhelmingly foreign. Much of this concern reflected the fact that American stations were easily accessible and those close to the border programmed (and advertised) with their Canadian neighbors in mind. According to Fortner (2005, p. 129), this question of "competing with the American broadcast networks and their affiliate stations near the Canadian border" was "one of the most significant issues raised" as Canada tried to develop its radio broadcasting industry and fight off fears of cultural annexation.

In the southern United States, cross border disputes were also being fueled in the 1930s with Mexican stations threatening to interfere with US radio broadcasters' signals. To complicate this matter was the fact that many of these stations were the result of US investors setting up in Mexico on the flat terrain in the border regions on the premise that American industries would advertise on such stations since Mexico was felt to offer "a fertile market for American products" ("New Mexican Border Stations," 1931, p. 10).

By the start of World War II, radio was firmly established as an important factor in international information sharing and in the ebb and flow of influences among nations. In times of hostilities (as in times of peace), radio could be a powerful tool for information, disinformation and subterfuge. Somerville (2012, p. 45) noted that the rise of the Nazis in Germany, the Fascists in Italy and the Soviet system all involved propaganda which had become "an acknowledged part of inter-state relations" and benefitted tremendously from the new technology since "the ability to broadcast across borders to promote ideologies, incite rebellion, national liberation or manipulate opinion had been vastly increased by radio."

RCA and NBC executive David Sarnoff (1939, p. 13) touted the democratic freedoms that radio in the United States enjoyed during the build-up to World War II, arguing that in some of the "autocracies if the Old World" broadcasting had been "converted into the most powerful instrument of dictatorship" such that certain governments were attempting to "tell their people which programs they may hear and which they may not," adding that "in some parts of Europe, to listen to a radio program originating in another country" was "to invite a jail sentence."

Several broadcasting operations developed and maintained targeted global operations before and after WWII. These radio stations, usually operating as government-run entities, later came to be known as "international radio broadcasters." During the twentieth century, and particularly after World War II and the popularization of national and commercial radio stations around the world, this method of international broadcasting became distinguished from local broadcasting, eventually becoming known as shortwave or "DX"ing (DX being short for distance). This remained a hobby for a small group of Americans, kept alive by enthusiasts through several decades and described in 1975 as "a minority habit" (Hale, 1975, p. xi).

While limited to hobbyists in the United States, in many other parts of the world, this system of international broadcasting remained important as a primary source of radio. So-called "international radio broadcasters" evolved into a global network of international news, information and propaganda in which competing messages spread from host countries to their colonies, former colonies, geo-political and military competitors, territories of interest and beyond. A 1954 media report ("Radio in the Cold War," 1954, p. 245) noted the vibrant role of these broadcasters in the Cold War and the geopolitics of the day in which governments spent "vast sums of money" to keep up "an incessant flow of news, comment, and plain propaganda from hundreds of broadcasting stations all over the world" representing "one of the major fronts of the cold war."

Among the operations comprising the international radio broadcasting system were long-established names such as the World Service of the British Broadcasting Corporation (BBC), the Voice of America, Germany's Deutsche Welle and Radio France Internationale (RFI). At its peak, this loose assembly of international radio broadcasters numbered about 80 players (Browne, 1982). These myriad interests might be found on the shortwave band nightly jostling for primacy in the minds of diverse international audiences, meeting occasionally to sort out their numerous quarrels over scheduling, frequencies and power or to negotiate relay arrangements.

Less commonly mentioned among the ranks of these international broadcasters are the operations of some former colonies such as All India Radio, which, as we shall see, boasted a global audience. Strictly speaking, wartime propaganda stations, broadcasting from one distant country to another (sometimes pretending to be in the target country) could also be considered international broadcasters. Additionally, several stations in the interwar and post–World War II periods also became de-facto regional and global broadcasters who, in the absence of competing signals, would reach relatively distant listeners.

The established names in international radio broadcasting such as the BBC World Service and Voice of America were not, however, limited, occasional or accidental in their approach to global reach. One of the reasons so much attention has been paid to these broadcasters, particularly in their heyday between the end of World War II and the start of the satellite era, is that they were deliberate arms of foreign policy as well as important sources of international news and information. Ionospheric bouncing and a system of repeaters throughout the globe served to amplify these broadcasters around the world on the shortwave band. These broadcasters also routinely translated their programming into multiple languages ensuring their messages reached broader audiences.

Fiedler and Frère (2016) noted that international broadcasters have been perceived variously as tools of Western imperialism, voices of public diplomacy and even as welcome sources of information (particularly where local broadcasting is heavily controlled or difficult to receive). These broadcasters have evolved through periods of global strife and political change, often as competing voices on the international stage. In a time, today, when almost any radio broadcaster can become effectively global, the distinction of being an international broadcaster is a dubious one, but in their own histories, each of these international broadcasters served as important tools in national expression for various groups. International broadcasters have also filled information gaps when national and regional communication media are disrupted or disabled such as during times of war and conflict.

Alongside relatively mundane social roles such as providing an international component to newscasts in many countries, international broadcasters would always be lightning rods for suspicion of propaganda. From subtle biases to outright false information, their opponents would accuse these global radio stations of not just keeping their citizens or colonies informed but also trying to influence the geopolitical situation. At no time were such charges more frequent and intense than during wars and other conflicts. In 1939, for example, *The Times of London* (p. 6) reported that the problems raised by the political use of the radio had "entered into everyday discussion" and included the difficulties of distinguishing real broadcasts from propaganda.

In describing the situation, *The Times* also cited a study from the Geneva Research Center arguing (perhaps not surprisingly) that their own British BBC was less guilty of propaganda than competing European international broadcasters (here called "intercontinental broadcasters"):

> It is remarked that the Corporation, though believed occasionally to eliminate information on grounds of expediency, did not exclude facts merely because they might be prejudicial to Empire prestige and even sought out contradictory interpretations of events, holding that as many different aspects as possible of any problem should be freely aired. ("Radio Politics," 1939, p. 6)

The Times provided a description of the pre-war and wartime strategies of the various global broadcasters whose signals were all capable of reaching each other, which included Russian broadcasts in over 62 languages denouncing Nazi and Fascist doctrines and the German Nazi regime, in turn, jamming their signals:

> The Nazis habitually jammed the Moscow station, and in a nightly program entitled *Here Speaks Moscow* clippings from Soviet newspapers, exposing the less savory aspects of conditions in the Union, were read and furnished with appropriate comments. Particular attention was called to the names of Jewish or supposedly Jewish officials in Russia. After every name the speaker emitted a loud, raucous, and sometimes melodramatic, "Ha, ha!" . . . Programs sent out daily to all sections of the globe, in five or more languages, were distinguished by the excellence of the music relayed as a seductive introduction to political talk. ("Radio Politics," 1939, p. 6)

The report ("Radio Politics," 1939, p. 6) described Italian shortwave broadcasts as "psychologically, more adroit" than their competitors with broadcasts in English and French, for example, that presented slightly different perspectives for their intended audiences. The Italian broadcasts would also include lessons in Italian language in which Mussolini's speeches "were read out as texts for the dictation" ("Radio Politics," 1939, p. 6). Concerning French broadcasts, the Geneva Research Center noted that the French provided broadcasts in Russian, Italian, Spanish, Portuguese, German and English that while "objective and well written . . . did not always include late events" ("Radio Politics," 1939, p. 6). However, *The Times* also noted that the effect of this French programming was diminished by the fact that the French shortwave station Paris Mondial was also subject to Nazi interference from the nearby, more powerful Zeesen shortwave station in Germany (later to become international broadcaster Deutsche Welle).

FOUR

Major International Radio Broadcasters

The following sections provide brief outlines of the development of several important dedicated international radio broadcasters who have acted as public voices of their respective nations or groups. The selection is not meant to be exhaustive, but rather a sampling of some of the operations that became well-known under this general label and exemplified the formal international broadcasting venture. The international radio broadcasters described below appear in roughly chronological order of their international service launch, bearing in mind that launch dates are often complicated as stations take different forms or names (or even fall under different control) at different times.

THE BRITISH BROADCASTING CORPORATION (BBC) AND ITS WORLD SERVICE

The Wireless Age (a Marconi company publication) noted in 1922 that the people of Great Britain were first able to "listen eagerly to their own broadcast programs":

> Broadcasting has begun in England, where four stations now are in operation. The public is listening with keen interest, and all reports from England show that our British cousins are going through just the same period of excitement over the new science that America experienced over a year ago. . . . The British broadcasting plan entails the licensing not only of transmitting apparatus, as in this country, but also of receivers . . . heavy penalties are prescribed for the use of a receiving station without this license. ("This is Station 2LO," p. 27)

The British government was unable undertake broadcasting on its own and started the British Broadcasting Company as a multi-business private venture on October 18, 1922 (Briggs, 1995). Several private broadcasters including Marconi had already been broadcasting with varying regularity at least two years prior but the British government sought to prevent the rapid spread of private broadcasters on multiple frequencies as had initially happened in the United States. A report in *The Cologne Post* in September of 1922 ("Broadcasting: Early Start," p. 4) described the emergent company as "an affiliation of all the principal companies which manufacture wireless sets" who were "affiliated for the reason that if each was to broadcast independently inextricable confusion would result."

In keeping with the global span of their empire in the early 1920s, British authorities cobbled together all existing broadcasting and transmission resources to support at least the image of global reach. Thus when King George V was scheduled to address the opening of the British Empire Exhibition in 1924, Associated Press reports carried British aspirations of their leader's voice being heard around the world:

> King George will "speak a piece" that is expected to be heard around the world at 11:30 A.M. Greenwich time, April 23, when he will formally open the British Empire Exhibition at Wembley. For the first time in English history, the actual voice of a monarch will be broadcast and heard simultaneously in the homes of hundreds of thousands of his subjects. ("King George's Voice," 1924, p. E1)

This grand attempt to have the king's voice span the globe enjoyed only limited success with radio enthusiasts throughout the empire reporting poor or no reception. Authorities circulated a text version of the speech by wired and wireless telegraphy as a backup measure. According to newspaper reports, "the royal voice itself, however, did not carry beyond the British Isles" as "radio fans in Canada, Africa, Australia and India listened vainly, except for isolated cases, when snatches of speech or music were heard" ("British Again Fail," 1924, p. J14).

The British Broadcasting Company license was a temporary one, extending only to 1926 and the government's uneasy relationship with "private" broadcasting was evident in continuing government investigations and reports regarding the future of broadcasting. In April 1923, the postmaster general appointed a committee to consider the state of broadcasting and to make recommendations about its future in Britain under Major-General Sir Frederick Sykes. The Sykes Committee reported on August 23, 1923, anticipating international broadcasting and affirming the notion of state control:

> In view of the possibility that all large communities may eventually demand this inexpensive service, and that Imperial and international broadcasting services may eventually be established, the Committee

considers that "the control of such a potential power over public opinion and the life of the nation ought to remain with the State." ("Report," 1923, p. 558)

The later Lord Crawford Committee reported to the British Parliament on
March 3, 1926, and also affirmed government control. Arising out of the
work of this committee, the British Broadcasting Corporation was established by royal charter on December 20, 1926. London's *The Economist*
magazine supported the public corporation since, it argued, "American
experience" had proven "beyond the shadow of doubt that broadcasting
must be a monopoly" ("The Finance and Future," 1926, p. 504).

In July of 1926, the British postmaster general, Sir W. Mitchell Thom-
son, announced in the House of Commons, a plan to establish the British
Broadcasting Corporation to take over the British Broadcasting Company
("British Broadcasting Corporation," 1926, p. 27). The transition from the
British Broadcasting Company to the British Broadcasting Corporation
took place on December 31, 1926. The corporation transmitted over
68,000 hours in 1927 with entertainment forming the bulk of all programs
("The British Broadcasting Corporation," 1928, p. 3).

The new public corporation met with widespread acclaim; radio
broadcasting as a national resource was the predominant sentiment, particularly since public memory of World War I was strong enough to
bolster acceptance of government control of what could be a strategic and
defensive asset. Despite this sentiment, however, the corporation was
still frequently criticized. The public accused the BBC of faults ranging
from poor timing of programs to even "too high morals" as one newspaper reported in February of 1927 when "radio programs of [the] new
British Broadcasting Corporation" were facing criticism from the British
public, who feared that "radio may soon fall into hands of pedants and
uplifters" ("Radio Suspected," p. 5)—pedants being a pejorative for overzealous instructional speakers and uplifters being religious preachers.

While touting the new BBC's promise of better news, educational,
sports and musical programs, *The Sunday Times* of London also highlighted the global (empire-wide) potential of the BBC noting that:

> Plans have also been prepared for the creation, probably at Daventry,
> of a special shortwave station for radiating British programs through-
> out the Empire. . . . Among the proposals to be made to the Corpora-
> tion is one which envisages the erection of high-power stations and
> listening-posts which would link the whole Empire together for broad-
> casting purposes. ("Better Wireless Programmes [sic]," 1927, p. 14)

As Berg (2013, p. 50) put it, "with the imperatives of empire had come a
new mission for shortwave." From its very first days, the notion of broadcasting to the distant corners of the empire was a tantalizing, if elusive,
objective. The BBC started experiments toward such distant broadcasts in
late 1927 with a shortwave transmitter at Chelmsford about 30 miles

northeast of London built in association with the Marconi Wireless Telegraph Company ("Britain's Broadcasting Service," 1935).

The BBC, in its 1928 report to parliament, noted that "although so far
no guarantee of service can be given, and undue expectations, of Empire
Broadcasting for the near future are to be discouraged, there is little
doubt that the prospects are encouraging" (The British Broadcasting Corporation, 1928, p. 10). Despite this cautious limiting of expectations, their
success in reaching the empire was evident when, in 1928, the BBC
broadcast news about the faltering health of King George V "from the
Chelmsford Short Wave station 5SW, for reception throughout the Empire" (The British Broadcasting Corporation, 1929, p. 10). The BBC later
presented proposals for what it called its "Empire broadcasting scheme"
but found inadequate support at the British and colonial imperial conferences in 1930 when the overseas colonies were less than enthusiastic
("Empire Broadcasting," 1933) whether for administrative or financial
reasons, or owing to the less than positive experiences with the earlier
Imperial Wireless Chain.

The racialized character of the British Empire and its broadcasting
was evident in the plans for an empire-wide broadcasting service. As
Potter (2008, p. 475) has noted, the British Broadcasting Corporation proposal for "Empire Broadcasting" submitted to the Imperial Conference of
1930 specifically referred to the scheme as being intended to reach the
"white population under the British flag." This racialized character of
British broadcasting enterprises was also evident in individual colonies.
Wilson-Heath (1986, p. 51) explained that the British introduced radio
broadcasting in Kenya in the service of colonists with a station in Nairobi
in 1928, noting that "Kenyaradio was primarily intended to entertain its
listeners and to provide a cultural link between widely scattered European homesteads and missions and with Britain." Wilson-Heath (1986,
pp. 56–57) explained the context of such early colonial efforts in Africa:

> Like the service in South Africa and those established in the next
> decade in Salisbury, Southern Rhodesia (1932), Lorenco Marques, Mo-
> zambique (1933), and Dakar, Senegal (1939), Kenyaradio was intended
> to serve the white community in the colony. Like the other early coloni-
> al services, it broadcast in the language of the colonial power and of-
> fered chiefly entertaining fare.

In 1932, having failed to gain financial support from the overseas
colonies, the BBC announced its decision to embark on the Empire Broadcasting scheme independently and to establish a special-purpose long-
distance broadcast facility at Daventry, citing increasing demand for the
service from overseas listeners (The British Broadcasting Corporation,
1932). The "Empire Service" began broadcasting from Daventry in December 1932 (Webb, 2014) with the empire divided up into five geographic zones and the station broadcasting directly to each of the zones.

From its start in 1932 to May of 1934, the Empire Service received over 35,000 reception reports and letters about its programming, and by 1934 the colonies were beginning to offer money toward the broadcasting effort ("Britain's Broadcasting Service," 1935).

The Empire Service was designed, wrote Wood (1992, p. 36), "with the purpose of keeping Britain's colonial civil servants in touch with the mother country." The service would broadcast news and entertainment to the colonies; distribution would include relays of broadcasts on local medium wave stations already in operations in many of the territories ("Empire Broadcasting," 1933). With the launch of the service in 1932, King George V took to the airwaves to greet the empire at Christmas. Wood (1992, p. 36) noted that, following testing at the Daventry short-wave station in mid-November, "on 25 December 1932 King George V became the first monarch to speak to his peoples throughout the world." The overseas focus changed in the following years, particularly as World War II influenced coverage and programming. As Webb (2014, p. 1) has noted, "transmissions to Europe began in September 1938 with German, Italian and French news broadcasts started at the time of the Munich crisis, in a deliberate attempt to counter the propaganda of Italian and German radio stations."

The global influences of the BBC, its Empire Service and World Service broadcasting were not solely one-way. While much of the programming contained strong British and Euro-centric cultural biases, there were also several efforts to include programming that was culturally relevant to the colonies. In 1939, for example, the BBC first broadcast their program entitled *Calling the West Indies* and in 1943 they started a follow-up to that program entitled *Caribbean Voices* to address audiences in that region and to highlight the issues that immigrants to England from that part of the world were facing (Newton, 2008). Later, this radio forum highlighted emerging literary talent from the Anglophone Caribbean. BBC producers were also known to collect recordings of performances from various colonies for programs that demonstrated the range of music and song throughout the empire.

Potter (2012) has argued that the BBC made some effort at pursuing a reciprocal flow of programming, particularly with the so-called "dominion" broadcasting authorities in the semi-autonomous territories which included Canada, Australia and New Zealand among others who were producing their own radio programs. The level of exchange, however, was never to make more than a passing impact on the primary flow of materials from the imperial capital. Another dimension of reciprocity may be found in the efforts of Sydney station 2FC and its operator Amalgamated Wireless who issued their own "Empire Broadcast" on September 5, 1927, with a "global ambition" and the notion that there was no reason for England to be the only source of such broadcasts (Given, 2009).

In the decades following World War II, the British Empire took on a new dynamic, with increasing pressure from their colonies for independence and the burdens of empire growing heavier on a war-depleted Great Britain. In that context, the BBC evolved into a major source of news and information across the globe. Its external and global character were evident in its name change to the BBC Overseas Service in 1939 and then to the BBC World Service in 1965, at which time the service boasted some 160 rebroadcasts of its sixteen main news bulletins (The British Broadcasting Corporation, 1966, p. 72). In many cases these rebroadcasts were a mainstay of news in former British colonies even when their local broadcasting had become relatively highly developed.

Germany's Zeesen, Deutsche Welle

The German push to establish radio began during tumultuous times with the establishment of its first formal station, the *Funk Stunde A.G.* in Berlin on October 29, 1923, during a time of hyperinflation and social unrest ("Broadcasting in Germany," 1927). Under dismal economic circumstances only about a thousand Germans could afford to pay the licensing fee of "60 gold marks or 780 milliards," a figure which rose to about two million by 1927 with the license fee set at 24 gold marks once the economy had stabilized ("Broadcasting in Germany," 1927, p. 6).

The operation that eventually became Germany's dedicated global broadcaster was announced in 1927 with plans to start broadcasting in December of that year from facilities at the Zeesen station. The history of Germany's use of radio and its global dimensions cannot, however, be separated from the looming presence of the Nazi regime and the role of domestic and foreign propaganda. The German homeland, Germany's neighbors and lands far afield would feel the impact of Nazi doctrine through radio well before Germany became a part of the relatively peaceful system of global broadcasting.

Graves (1941, pp. 13–14) placed the start of Germany's international broadcasts in 1933 about a month after Hitler came to power, ostensibly to promote trade ("so that many talks were given about the excellence of German workmanship and the quality of German goods") and to "establish a unity of thought among Germans all over the world" ("even broadcasts to the United States urged German-Americans to 'remember the Fatherland'"). However, before long, the real purpose of German international radio broadcasting under the Nazi regime—that of assisting in "the expansion of German power" (Graves, 1941, p. 14)—became clear:

> German programs were broadcast in 1933 for the purpose of inciting revolution in Austria and thus making that country an easy prey for the Reich. Radio broadcasts were extensively used in 1935 to persuade inhabitants of the Saar, separated from Germany by the Versailles trea-

ty, that they should vote to join the Reich again; and radio was used to prepare the stage for every succeeding German expansion.

Hale (1975, p. 1) wrote that when the Nazi Party came to power in 1933 radio was still "a novelty" whose "propaganda potential was virtually untried" but that "the Nazi leaders' faith in it was unbounded." Wood (1992, p. 40) emphasized that Hitler recognized radio broadcasting as an important element of strategic communications and noted that the regime undertook construction of shortwave stations at key locations including Zeesen such that "(shortwave) transmitters were beaming the voice of Nazi Germany to all parts of the world."

Nazi leadership considered radio such an important tool for instilling uniformity among the German people that independently operating radio stations in German cities were brought under centralized Nazi control in 1933 (Rhodes, 1976). Gower (1939, p. 294) argued that this move converted German wireless "into part of the huge state propaganda-machine." *All-Wave Radio* ("Foreign News," 1936, p. 128) noted that German radio reflected the racist and nationalist policies of the ruling party:

> Jewish music is absolutely banned, and Negro so-called "jazz" in the same category—modern dance rhythm is considered moronic and degrading. German military music, Bavarian peasant tunes and Tyrolean yodels are etherized 99.9 percent of the time stations are on the air.

The regime also made available a subsidized cheap radio receiver known as the *Volksempfänger* (People's Radio or People's Receiver) sold at about a week's wages among the working classes. According to König (2003), entertainment content captivated listeners and made them more amenable to the political content. Cheap receivers were also necessary since radio ownership was quite low in 1933 prior to the introduction of the *Volksempfänger* with only about one in four German households overall owning a set ("News Notes," 1931) and only about one in ten families among the working classes being able to afford one.

German radio technicians would "practically force the sale" of the *Volksempfänger*, which was designed to make it "almost impossible to pick-up non-German programs" ("Foreign News," 1936, p. 128). By 1939 over half of the population had access to a radio in the home (Gower, 1939, p. 294). An interviewee in Birdsall's research commented (2012, p. 11) that "the state-subsidized radio sets (*Volksempfänger*) had the purpose of keeping the people acoustically under control." This control was also achieved through what US prosecutors later described as "an elaborate network of closely knit and exclusive organizations of selected volunteers oath-bound to execute, without delay and without question, the commands of the Nazi leaders" that exercised "drastic controls over German society" (Office of United States Chief of Counsel for Prosecution of Axis Criminality, 1946, pp. 1–2).

To further bolster listening within Germany (and foster their hold on the audiences), the authorities "encouraged" communal listening:

> When a speech by a Nazi leader or an important announcement was to be made, factories and offices had to stop work so that everyone could listen. . . . Another technique used was the "radio warden" for each block of houses or apartment buildings. This Party member would encourage his neighbors who did not own a radio to buy one (sometimes he would lend them the money to do so); otherwise, to listen to important speeches in his or a friend's home. He sent in regular reports on their reactions to the broadcasts. (Rhodes, 1976, p. 27)

Rhodes (1976, p. 27) also noted that this social control extended to foreign listening since "the radio warden became of special important during the war when he reported those listening to foreign broadcasts." Such listening would have likely required listening apparatus other than the *Volksempfänger*, and Gower (1939, p. 295) suggested that German listeners with appropriate listening devices (many of whom spoke other languages) were fond of listening to broadcasts from Great Britain.

Under the Nazi regime and centralized control, the German radio system grew to include 38 medium and long-wave broadcasting stations in 1938, and Gower (1939, p. 294) observed that the "frankly propagandistic" nature of this "immense network" under the almost singular control of Goebbels had global aspirations:

> Whereas in 1933 Germany only possessed three relatively small short-wave stations, she now has fifteen very powerful ones and the number of hours per day for broadcasting has increased from 2 in 1933 to 109 in the current year (1939). These transmitters broadcast to six different zones (Africa, East Asia, South America, South Asia, Canada and North America). Although Germany has no oversea possessions, there are 15,000,000 Germans, or descendants of Germans, living abroad. German officials in an interview with World Radio declared that this service was intended as a link between these people and the mother country.

Gower noted that the German network featured broadcasts for schools that reached 35,000 schools every day. These broadcasts were known as the *Deutsche Welle*, or the German Wave, the name that was applied to Germany's official global broadcaster eight years after the end of World War II in a partitioned Germany (Deutsche Welle, 1982). Kirch (1968, p. 117) described the founding of *Deutsche Welle* in the Federal Republic of Germany and its early evolution:

> In 1953 the West German Regional Broadcasting Stations got together to set up a shortwave station under this name for the purpose of providing listeners abroad with a true picture of the cultural and political life of present-day Germany. At first the broadcasts were made only in the German language, but so many letters were received in other lan-

guages, that it was decided to add programs in English, French, Spanish and Portuguese.

By 1967 *Deutsche Welle* was even broadcasting programs to Africa in multiple languages amounting to some 17 hours daily, surpassing Voice of America programming in that region ("Germans in Africa," 1967, p. 11). Despite the negative associations of German international broadcasting in and around World War II, Germany's international broadcaster was still capable of spreading goodwill about Germany and furthering German diplomatic and hegemonic influences in other nations (Deutsche Welle, 1982).

Marconi's Radio Vatican (Radio Vaticana)

As early as 1923, the Roman Catholic Church expressed interest in using (audio) radio to reach global audiences ("Churches to Use Radio," 1923). Though the Vatican launched *Radio Vaticana* in 1930, radio magazines such as *The Wireless Age* and *The Oscillator* reported murmurings about plans for an antenna on the dome of St. Peter's from as early as 1918. Such a service, argued the wireless enthusiasts, could provide not only a useful conduit for transmission of war messages but also a potential news service with global reach—particularly since the Vatican was connected to church interests around the world. In discussions with Marconi and representatives of the Radio Corporation of America, then newly appointed Pope Pius XI made clear his interest in the long-distance capabilities of the radio facilities he hoped to establish ("Says Radio Device," 1924). Saerchinger (1938, p. 67) noted that the radio station was established "so that the Pontiff might communicate, directly and independently . . . with his Nuncios throughout the world" and with the close cooperation of the then Senator Guglielmo Marconi who had been "raised to the rank of Marchese by the King of Italy" and had become "one of the chief ornaments of the Fascist State, an elder statesman and president of the Italian Academy." Saerchinger (1938, p. 67) also noted that Marconi had married into a noble Italian family and was a "confidant of the Pope."

Darrah (1929, p. 18) noted another international dimension to the Vatican's radio aspirations in that American money and expertise were partly behind the proposed radio station:

> The station will also be built by an American Company and paid for by American Catholics. It will cost 30,000,000 lire (about $1,575,000). . . . Broadcasting will be a feature of this station, enabling the Pope to disseminate his speeches and communications among 400,000,000 Catholics throughout the world.

The primary funding for the Vatican's radio station came from the Italian government and the Vatican. Saerchinger (1938, pp. 67–68) clarified, however, that the vision of the Pope's radio station was somewhat more luxurious than those funds might have allowed and that the shortfall was covered by American donations:

> What was wanted was not only a very modern and powerful station, such as the supreme pontiff should command, but a marble building and the most sumptuous accessories. . . . The difference between the available funds and the actual cost of the proposed Vatican station was, it is said, contributed in part by donations from the Faithful (chiefly in America).

The proposed launch of Vatican Radio was announced in 1930 and the handover to the Pope happened on September 22nd of that year. The formal launch of the station with addresses by both the Pope and Marconi took place on February 12, 1931, after successful tests received as far away as New York City. The Pontiff's speech on that day was entitled "Wireless Message of His Holiness Pope Pius XI Broadcast to the Whole World" in which the Holy Father referred to himself as "the first Pope to make use of this truly wonderful invention of wireless" (Ratti, 1931, p. 1) and addressed "all ye nations," giving glory to God who, in Pius' words, "in our days hath given such power to men, that their words should reach in very truth to the ends of the earth" (Ratti, 1931, p. 1). Adler (2004, p. 399) noted the importance of the multilingual and international focus of the Vatican station, observing that by 1934 operators were charged with "broadcasting in a variety of languages."

Italian international broadcasting efforts outside Radio Vatican took less of an episcopal or evangelical character and embraced, instead, the tenets of Fascism. While Radio Vatican continued its existence, secular Italian international broadcasts also began to reach out to international populations. According to Graves (1941, p. 14), starting in 1935, the Italians embarked on a program similar to that of Germany, using radio as a tool of political propaganda:

> Particular effort was made, for example, to instill in the large Italian population of South America the ideas of Fascism. In 1937, Italian broadcasts in Arabic to Palestine and the Near East were of such a revolutionary character that Britain protested.

Over its years of existence, Vatican Radio has encountered praise for its role in such undertakings as promoting religious freedom, particularly as resistance against communist repression in Eastern Europe and the former Soviet Union ("Vatican Radio," 1991). The station has also, however, encountered criticism for some of its actions (or inactions) including its perceived lack of response to the atrocities in Nazi Germany against the Jewish community (Adler, 2004) both in Europe and abroad.

In support of this latter perception Cianfarra (1944, p. 7), suggested that in and around 1942, audiences in the United States felt that then

Pope Pius XII was not using Vatican Radio to clearly speak out against the Nazi movement and campaign in Europe and, rather, "was sitting on the fence watching which way the war was going while preparing to join the victorious nations." "In other words," wrote Cianfarra (1944, p. 7), "there was a deep-rooted suspicion that the Vatican was playing both sides in the war, and that it had not taken a firm, unswerving stand."

Radio Vaticana has continued to broadcast over the air and added other transmission modes. In 2004 the station boasted broadcasts in 34 languages utilizing satellite, short wave, medium wave and frequency modulation (FM) and was transitioning to include digital terrestrial broadcast as well ("Spotlight: Vatican Radio," 2004).

Radio France International

During the late 1920s, the French, like other European powers of the time, sought to control the use of radio within and around their borders. For example, during November 1924 the French Ministry of the Interior used a special system that monitored amateur radio conversations and broadcasts:

> Radio, which has gained more than 1 million native devotees in France, will be subjected to police inspection and control. To this end, a receiving station is being installed at the Sureté General where a special service has been organized to "listen in" and censor future air conversations and messages. The reason for this movement, for which the Minister of the Interior is responsible, is the possibility which exists at present that the enemies of France may use the radio to spread harmful propaganda. ("Police of France," 1924, p. 2)

French radio broadcasting officially started with the French National Office of Radio Broadcasting in 1929. Their commitment to a non-commercial approach to broadcasting was evident in the announcement in December 1934 that all advertising was banned from French radio stations on the authority of the French minister of posts, telegraph and telephones ("France Bans Advertising," 1934):

> Among the French radio stations of that time *Poste Colonial* was special in its global focus. Like other colonial powers, France also sought to harness radio to suit its imperial purposes. According to Graves (1941, p. 13), France "during her Colonial Exposition in 1931 began to broadcast programs to her Empire—speaking, however, only French and the languages which were spoken in French colonies" and with the purpose of "keeping the loyalty of her overseas colonies, and in maintaining the *status quo*." In pursuit of maintaining its Empire, French authorities established *Le Poste Colonial* (Winston, 1998) to broadcast to their colonies. This service was renamed to *Paris-Mondial* in 1938 until World War II put an end to its broadcasts in 1940.

The French found their international broadcasting efforts diverted from their colonies to more immediate threats from their own region. In 1936, in the face of German threats, according to Graves (1941, p. 14), France started "an aggressive foreign-language program, broadcasting in German, ostensibly for the benefit of Alsatians who spoke German, but actually for Germans themselves."

Graves (1941, pp. 19–21) provided an account of *Paris-Mondial's* program of broadcasts to the United States that, at times, ran for six and a half hours per day starting in the spring of 1938 and how this programming changed with the emerging geopolitical situation, particularly after Germany invaded Norway and Denmark in April 1940:

> News and comment, meant to persuade American listeners to hate Germany and to have confidence in France, now took first place. . . . Documentaries designed to bring the war home to Americans received new emphasis. . . . On June 3, for example, recordings were played so that Americans could hear the sound of bombs falling on Paris. News and comment were now not only engaged in the mission of inspiring confidence in France and hate for Germany, but in asking America for help as well.

When the French surrendered to the German invasion, *Paris-Mondial* fell under German control, and under the terms of the Armistice of June 22, 1940, there were to be no more French international broadcasts. German authorities commandeered the transmitters and began to use them for German broadcasts instead.

Many years later, in 1975, France resumed dedicated international broadcasts with *Radio France Internationale* (RFI). Its eventual goal was to launch a full-time global broadcast service in French modeled on the BBC (Bradley, 1981). In 1983, RFI was still being described as a "small voice" ("Closed Circuit," 1983) but announced plans for increasing its programming from 125 hours weekly to 740 hours weekly as the French government proposed an increase in the station's budget from 149 million francs in 1982 to 390 million in 1986. According to Wood (2000, p. 63), who estimated RFI's listening audience in the 1990s at about 80 million people worldwide, the station's program output "is directed to France's former colonies along with other French-speaking countries in the Caribbean and parts of the Pacific."

All India Radio

Precursors

With a land area many times greater than most European nations and a varied linguistic landscape, India faced numerous challenges to workable radio. Among the first examples of wireless audio broadcasting was an August 1921 broadcast of music from Bombay organized by the *Times of India* newspaper and the Posts and Telegraph Department at the request of Sir George Lloyd, then governor of Bombay, who listened from some 175 kilometers away at Pune (Luthra, 1986). Discussions about broadcast stations and government rules for running them were popular as early as 1922 (Banerjea, 1922). By 1923, the Indian government had "decided to permit private enterprise to undertake broadcasting in British India" and scheduled for the 7th of March that year a conference that would bring together British and Indian equipment manufacturers together with the director general of Posts and Telegraphs at Delhi with a view to facilitating the initiation of radio broadcasting ("Broadcasting in India," 1923, p. 13). BBC Director General John Reith had also shown an interest in Indian radio around the same time but conceived of a London-based operation for the scheme, which failed to impress either the British colonial India Office or the viceroy of India (Luthra, 1986).

In 1922 the director general of Posts and Telegraphs, Lieutenant Colonel Hubert A. Sams, successfully proposed to the Indian colonial government establishment of radio broadcasting that would generate revenue through license fees for transmitters and for receiving sets. At the 1923 Delhi Broadcasting Conference, Sams told manufacturers and the press that the Indian government had received "about 18" requests from private firms for licenses to broadcast but argued that "it was quite impossible for Government to permit broadcasting by individual firms" and had decided that "if on proper conditions and under reasonable control broadcasting was to be permitted in India, it could be permitted only as in the United Kingdom through a single licensed company for the whole of India ("Broadcasting: The Delhi Conference," 1923, p. 10). The conference produced more questions than answers, however, and left uncertainty about how the emerging broadcast enterprise would unfold.

Numerous demonstrations of wireless audio broadcasting around India (some of which bore the mark of Marconi's public relations) also fueled interest in radio. Banerjea (1923, p. 77) noted that the demonstrations were "being conducted by the Indian States and Eastern Agency, who are the sole agents for the Marconi Wireless Telegraph Company." Marconi also promoted interest in radio through loans of low-powered transmitters to so-called "radio clubs" in cities including Bombay, Bengal and Madras ("Echoes and Memories," 1937). Baruah (1983, p. 1) credited these clubs with the genesis of radio in India and noted that "the Radio Club of Bombay broadcast its first programme in June 1923, and the Calcutta Radio Club in November 1923 with transmitters loaned by the Marconi Company." A year later, on July 31, 1924, the Madras Radio Club went on the air with a 40-watt transmitter brought in and assembled from England. According to Baruah (1983, p. 1), "it was assembled by C. V. Krishnamurty Chetty who brought the components from England on completion of studies there" and the club broadcast "two and a half hours of transmission consisting of music and talks" every evening.

The first formal radio broadcasting station in India started in July of 1927 in Bombay as the Indian Broadcasting Company (IBC) with several radio clubs agglomerated under private business interests. The IBC launched a second station in Calcutta in August 1927. The government insisted the broadcasters should fund themselves from listening license fees (a portion of which went to the broadcasters when government collected the fee) (Ramasastri, 1959). The number of licensed listeners at the start was small (estimates vary widely from less than 1,000 to 3,600) in 1927, increasing after the introduction of broadcasts from the BBC to more than 16,000 by 1934 (All India Radio, 1940; "Broadcasting in India," 1940) and then to nearly 100,000 by 1939 ("News in Brief," 1939). With the demands of population, geography, and emerging concerns about security, colonial authorities took over broadcasting in 1930 with the formation of the Indian State Broadcasting Service (IBS) when the Indian Broadcasting Company went into financial collapse (Ramasastri, 1959).

After broadcasting took hold, the scope of programming available to listeners in India was quite wide. Local broadcasters in Bombay (VUB), Delhi (VUD) and Calcutta (VUC) provided news, music and talks in Hindi and in English. VUB Bombay provided in its first programming segment, at mid-day, programs entitled *Selection of Indian Music* at 12:00 noon and *European Lunch Hour Selections* at just after 1:00 p.m ("Radio Broadcasts," 1936). Relays of film music featured later in the day as were live performances of both Indian and European music. Records show additional content as a (perhaps unlikely) presentation of *Negro Spirituals* featuring "gramophone recordings" ("Radio Broadcasts," 1936). Calcutta's VUC offerings included live vocal and musical selections from artists in studio as well as transcriptions (i.e., recordings) of musical and variety pieces from the United States. Delhi's VUD, coming on air at 5:30 or 6:00 p.m., boasted a house orchestra called the IBS Orchestra under their conductor S. S. Niazee, who sometimes opened their programming day with traditional musical offerings such as *Durga Bilwal Thath*. VUD also featured *Ghazals* (songs derived from the romantic Persian poetry tradition) by a Miss Rajjo as well as *Thumris* (classical or semi-classical love songs) and *Bhajans* (religious songs) by Master Luchhi Ram. All these stations also featured newscasts in English and Hindi as well as various talks and lectures (Luthra, 1986).

As might also be expected, British Broadcasting Corporation programming accounted for a large portion of the radio diet of this (then) British colony. The BBC's many transmissions on several different frequencies included European classical music, news and other varied content. Offerings included musical presentations with titles such as *The BBC Dance Orchestra, Ambrose and His Embassy Club Orchestra, Pianoforte Solos* and *Military Band Music* alongside news, talks and such empire-focused programming as *Bells, and an Empire Service, Relayed from St. Paul's Cathedral, London* ("Radio Broadcasts," 1936). Additionally, transcripts of BBC programs formed a part of the broadcasts of the Indian stations in Delhi, Bombay and Calcutta.

Alongside these broadcasts of local and colonial scope, listeners could also receive broadcasts of a somewhat more foreign (and perhaps politically sensitive) nature. Indian radio listings of the time also featured programming broadcast directly from Germany in German, English and Dutch on several frequencies. These broadcasts included news in all three languages and programming with titles such as *Today in Germany — Sound Pictures*, *The Best German Male Choirs* and *Hitler Youth Programme* ("Radio Broadcasts," 1936).

From IBS to AIR

In 1936 authorities renamed the Indian State Broadcasting Service to All India Radio (AIR) after they had moved to integrate existing stations into a system that would reach the vast majority of Indians in India and also reach audiences abroad. The system, even domestically, faced several challenges—not the least of which was the fact that most villagers in rural areas could not afford to purchase receivers, prompting the need to develop sets and systems for communal listening.

The need for such communal receivers was already a matter for discussion in 1932 when the Indian Village Welfare Association (an organization of British colonials who aimed to improve life in remote and rural Indian communities) discussed the establishment of a "system of rural radio education and propaganda in India" ("Education by Radio," 1932, p. 10), describing the need for communal listening systems while also revealing perceptions of radio as a tool of social development.

In what the *Times of London* called an "ambitious program," India's foray into international broadcasting began as AIR, a service that would attempt to reach the entire subcontinent with its more than 200 dialects and, while offering international broadcasts to its population, also providing its content to listeners outside:

> The intention is to supply services in the vernacular for the entire country, allowing for the linguistic and cultural claims of each area; to make the short-wave services of Great Britain and Europe available to Indian listeners by relaying; and to interpret India to the world through the facilities which the short-wave system will offer to listeners in other countries. ("Broadcasting: The Community Set," 1937, p. 51)

Among these "other countries" AIR could count, by reception reports and official correspondences, were South Africa, Ceylon (Sri Lanka), Malaysia, British Guiana and Trinidad ("A.I.R. Programs Overseas," 1938). This network, with its capacity for covering great distances, became a weapon in the wartime arsenal of the British prior to Indian independence including expanded wartime services from AIR ("News in Brief," 1939). A special correspondent for the *Times of London* described the involvement of AIR with the British war effort in December of 1939 ("The Indian War Effort," p. 7), notably pointing to the inclusion of languages aimed at audiences outside of India as well as the relaying of London news. Singhal and Rogers (2001, p. 67) noted that during World War II "AIR launched an External Services Division and a Monitoring Service as part of the British Military Intelligence Wing" that broadcast anti-Nazi and anti-Japanese propaganda (often in local languages) into neighboring Asian countries. According to *Variety* magazine ("International: India's Radio," 1949, p. 13), this focus on the external impact of Indian broadcasting during wartime led to an expansion of AIR's global broadcasting reach when the British government "realized the potentiality of radio for propaganda" and moved from simple monitoring to active propaganda and counter-propaganda efforts.

Berg (2013, p. 240) noted that "the facilities of All India Radio were put to use to support Allied troop morale" with some focus on US troops including broadcasts "signing on with the 'Stars and Stripes Forever,'" and announcements such as, "This is the voice of the United States broadcasting to east from Delhi." These special program on AIR included re-broadcasts of content which Mackenzie (1999, p. 37) identified as being from the United States Armed Forces Radio services emanating from the War and Navy Departments of the United States and having as its slogan: "The Voice of Information and Education." The newspaper *C.B.I. Round-up* of February 4, 1943 ("Radio Programs," p. 2), reported that All India Radio carried programs including *Your Broadway and Mine* (featuring various Broadway performances) with episodes that included stars like Milton Berle and Phil Silvers, several episodes of *Yank Swing Session* (a 60-minute show that featured music and a guest DJ) and *Downbeat* (a program that highlighted American jazz music).

Britain's emphasis on bolstering the wartime and global presence of AIR was also, in part, a response to Axis efforts at propaganda broadcasts with station identifications such as "Free India Radio" and "Radio Himalaya." These Axis-backed broadcasts sought to destabilize Allied war efforts, sowing doubt about American intentions in the war and even suggesting, at times, that the Americans were after securing parts of the British Empire for themselves ("Axis Contradictions," 1942). The "Free India" radio station had the added global intrigue of being the work of an Indian nationalist in exile named Subhas Chandra Bose (once the mayor of Calcutta) who, with the help and support of the Nazi regime, broadcast his anti-British propaganda to India from Berlin (Hayes, 2011), Singapore (Rhodes, 1976) and other locations starting in March of 1942 (Berg, 2013). Rue (1942, p. 5) noted that "Free India" radio had been "telling the Indian people that they have everything to gain by allying themselves with the tripartite powers." Listeners in places as distant as St. Petersburg, Florida, reported picking up the "Voice of Free India" and transmissions under its alias "National Congress Radio" (Morrison, 1942).

Radio Himalaya was an effort by another Indian nationalist who hoped to capitalize on the war to displace British rule in India. Mohammad Iqbal Shedai (an Indian in exile who had engaged in subversive activities against British rule) was the man behind the Radio Himalaya broadcasts that pretended to be from northern India but were most likely from Italy (Prayer, 1991). Prayer (1991, p. 267) argued that the purpose of Radio Himalaya, which broadcast in Hindi and English, was "to increase the anti-British feeling among the people at a time when India's role in the war was becoming increasingly significant."

Outside of the wartime context, AIR broadcasts began to reach distant, and perhaps unintended, destinations. Their broadcasts of news and music were eventually being received in locations such as Trinidad & Tobago in the Caribbean and British Guiana in South America. In those places the descendants of Indian indentured laborers (many of whom would still have spoken or understood some amount of Hindi or other Indian languages during the 1940s) eagerly tuned in to AIR broadcasts and, when atmospheric conditions permitted, listened to distant voices brought near. AIR's global listenership, however, faced a sharp decline in the period after Indian independence in 1947 when the new Indian government sought to exclude mainstays such as popular film music from AIR on the grounds that it was not sufficiently reflective of Indian cultural values, a decision that would lose the service listeners abroad and at home. Singhal and Rogers (2001, p. 68) described how domestic listeners within India even turned to foreign sources such as Sri Lanka and (then Portuguese) Goa for their entertainment in response, noting as well that "only in 1957, 10 years after independence, did AIR launch a *Vividh Bharati* entertainment channel, broadcasting Indian film music and other entertainment fare." Rao (1986, p. 109) described AIR's external operations beaming to 55 countries in 17 foreign and 8 Indian languages during the 1980s.

The Voice of America

In the United States, radio broadcasting developed primarily as a commercial enterprise following government intervention in radio during World War I. Indeed, in the years following this early government intervention, there was some skepticism in the United States about government being allowed too much of a hand in the development of radio. Louis McFadden, a representative from Pennsylvania in the US House of Representatives, addressing the House Committee on Merchant Marine, Radio and Fisheries in March of 1934, argued that "the introduction of any resolution pertaining to broadcasting or the freedom of the air brings forth a great volume of interest from the American public" out of concern "that the control of the air is not abused, but that it is kept open and free for the benefit of all of the people of the United States" (United

States Congress, House Committee on Merchant Marine, Radio, and Fisheries, 1934, p. 18). Americans had been concerned for some time that radio under government control as in Great Britain could be used as a political weapon ("British Strike," 1926, p. 8).

This fear of government control and attendant concern about how radio might be influenced or manipulated was not unanimous sentiments. At about the same time, commentators also expressed concerns about how radio might survive without support and went as far as recommending that government become involved in forcing audiences to pay for listening. Raymond Francis Yates, a member of the Institute of Radio Engineers, argued for "a non-partisan Federal Commission of educators, entertainment and technical experts to govern, regulate, and control broadcasting and to arrange for the collection of a small yearly fee from each owner of a receiving set" (1924, p. 604).

Despite such suggestions, it became increasingly clear that the radio industry in the United States would not enjoy the benefits (or responsibilities) of government subventions or listening set license fee revenues as was the case in other places such as Great Britain and India. Thus, not long after radio broadcasting began to gain popularity in the United States, various attempts were being made to generate revenue from advertiser involvement.

Even for large powerful corporations such as AT&T and RCA, operating a radio broadcasting operation was not a feasible business operation without the possibility of generating revenue. The substantial costs of equipment and program production required that incomes be generated if broadcasting was to be sustained. The strategy for recouping the costs of radio broadcasting that emerged was that of selling portions of broadcast time to advertisers. Some informal commercial activity had been cropping up across early broadcasts when, for example, broadcasters would thank record stores who provided their music on the air. However, a formal system of payment for airtime mentions was first developed at AT&T's WEAF, which allowed program sponsorship starting in February 1922. Under such a scheme, a sole sponsor's name and product would be mentioned before and after a programming block for a fee. Lewis (1933, p. 17) described the typical appearance of a sponsor message at the end of a programming segment:

> From the radio set comes the closing strain of the orchestra. The announcer cuts in with "You have just heard a group of numbers by Madame L'Aria, by courtesy of — —" and follows with the name of the sponsor and the product advertised.

Intellectuals and audiences alike treated the advent of commercial messages with some suspicion. H. O. Davis, publisher of the Ventura Free Press, decried the evolution of such limited sponsorship messages, arguing that AT&T's WEAF precedent led to a broader onslaught of advertising practices on the radio, thus opening the gates for a "deluge of advertising ballyhoo" to pour "through the air into the American home" (1932, p. 43). Wood (1992, p. 27) noted that WEAF's decision to sell air time in 1922 "planted the seeds of a new business that eventually grew to envelop the broadcasting industry" that included advertising, public relations and propaganda.

In 1924 the US House of Representatives member from New York's 10th congressional district, Emanuel Celler, made publicly known his distaste for radio advertising, particularly of the kind that attempted to mask its presence and sought to support legislation outlawing the practice of such stealthy activity that he termed "indirect advertising" ("Celler Would Curb," 1924, p. 24).

Owing partly to the size of the United States and partly to the challenges that individual stations faced in creating content for a full broadcast day, the early radio industry emerged with a network model in which a group of stations would contribute and share content to fill the programming day. This model used wired telephone line technology that enabled reliable linkages among stations and expanded station reach. The pioneers of such linkages were AT&T (who controlled the telephone line technology) and the Radio Corporation of America (RCA), and the prominent networks that emerged were RCA's National Broadcasting Company (NBC), the Columbia Broadcasting System (CBS), the Mutual Broadcasting System and the American Broadcasting Corporation (ABC). Within these networks, listeners also had a choice of sub-networks such as NBC Red and NBC Blue and even NBC's Pacific Coast Gold and Pacific Coast Orange networks. Notably, NBC also had network stations in Montreal and Toronto, Canada (US Congress Senate Committee on Interstate Commerce, 1935, p. 4086).

The National Broadcasting Company (NBC) was incorporated on September 9, 1926, and owned several subsidiary companies (US Congress Senate Committee on Interstate Commerce, 1935). The oldest of the major networks, NBC developed a reputation for international content and a global scope early in its existence:

> The first scheduled short-wave broadcast, making possible international communication, did not come until 1929, and since then the National Broadcasting Company alone has relayed to audiences here more than two thousand programs from Europe. Crossing a lower threshold to the consciousness than the printed page, speaking an international language when it puts music on the air. (Bent, 1937, p. 118)

On New Year's Day 1931, one such broadcast featured Italian Premier Benito "*Il Duce*" Mussolini, who delivered a prepared address in English to the people of the United States via NBC by arrangement with the Hearst Organization (Barnouw, 1966, p. 250). In his speech Mussolini argued that Fascism posed no threat and that he and his country had no desire for war, while sending "greetings to the American people" and "warm feelings of friendship for their great republic" ("Italy Doesn't Itch for War," 1931).

NBC flexed its muscle in terms of coverage, equipment and innovation in radio broadcasting in many ways. One of its ventures to demonstrate its global reach and capabilities was involvement with the Holden (sometimes called the Terry-Holden) Expedition in 1937 into the jungles of British Guiana. NBC covered this scientific and photographic expedition through the establishment of a portable transmitter in the jungle and often with the help of the local Guianese station as a relay. The expedition provided weekly reports on their progress to listeners in the United States from October 1937 to January 1938 ("Jungle Radio," 2011; "Jungle Broadcasts," 1937).

The NBC network claimed a total of 4,818 international reports from 26 European capitals during a three-year period of World War II (1939–1941), having, as it recounted, "dispatched correspondents to every quarter of the globe, to every important world capital, to every war front":

> NBC microphones carried the war into American homes . . . schooled the American people in global thinking which did much to condition and prepare a peace-loving nation for its inevitable role as a fighting ally of the then-warring Democracies. (National Broadcasting Company, 1944, p. 50)

Even though NBCs primary mission was programming for the US audience and their foreign content formed only a small part of that programming, the geopolitics and business circumstances of the day also prompted program flows from the United States to other regions. Bent (1937, p. 119) described, for example, efforts to beam programming to Latin America, eclipsing earlier European dominance in programming to that region with moves such as providing "complete coverage of the Inter-American Conference for the Maintenance of Peace, at Buenos Aires" and a South American tour by an NBC vice president named John F. Royal, who talked with "presidents and other high officials" and arranged to "send at least six broadcasts weekly."

CBS made similar claims to bringing international news and notable international figures to the American radio audience and to taking US broadcasting to global audiences. Notably through the work of its correspondent César Saerchinger from the early 1930s, CBS secured broadcast addresses from figures such as Mahatma Ghandi, Leon Trotsky and Pope Pius XI (Barnouw, 1966; Saerchinger, 1938). CBS also made efforts at reaching international audiences with efforts such as a 1932 "program of American talent" called *Hello Europe* broadcast in the United States and relayed throughout Europe in partnership with European counterparts including the BBC and stations in Germany, France, Italy, Czechoslovakia and Hungary ("Reception of CBS," 1932, p. 15).

The US government was aware of the potentials of broadcasting during both active hostilities and simmering Cold War tensions. As early as 1938, the House Committee on Naval Affairs was making efforts to establish a government radio broadcasting station that would provide educational content and "programs of national and international interest" (House Committee on Naval Affairs, 1938, p. 3469). In that same year, President Roosevelt appointed an Interdepartmental Committee on International Broadcasting, and since that committee was yet to report, the naval brass were stymied in their efforts to move ahead with their station. As the pressures of World War II moved from foreign policy positions to military ones, a more pragmatic approach to radio and international broadcasting had to be adopted.

An example of the wartime radio propaganda efforts of the United States was the ABSIE, or the American Broadcasting Station in Europe, established through a cooperative venture between the US Office of War Information and the BBC in order to "bring the voice of America more powerfully to bear in the struggle against the enemy" (The British Broadcasting Corporation, 1947, p. 29). This was a short-lived effort, starting in April of 1944 and ending on July 4, 1945, from studios in London. At home, in part to address the home audience, but also with an awareness of the great international reach of US radio broadcasters both directly and through relays, US military authorities produced programming that was distributed to stations who were encouraged to broadcast the content in support of the war effort.

Abroad, shortly after the ABSIE experiment, US war efforts at radio broadcasting began to take more concrete shape, with the notion of the "Voice of America" being central to the strategy. A US government report from 1946 (United States Bureau of the Budget, p. 218) noted that the government had to act to "combat the insidious influence of Axis propaganda." At the start of US engagement in World War II, authority for wartime radio broadcasts lay in the hands of the Office of the Coordinator of Inter-American Affairs and the Office of the Coordinator of Information, which worked with "the State, War, and Navy Departments" (United States Bureau of the Budget, 1946, p. 218). Soon after, the US military authorities created the Foreign Information Service, which "undertook to spread the gospel of democracy" and through which "radio programs were presented as 'The Voice of America'" (United States Bureau of the Budget, 1946, p. 219).

These efforts eventually gave rise to a unified system of US foreign broadcasting called "The Voice of America" that, in cooperation with the BBC and British authorities, added the American voice to the war. In 1948, legislation entitled "United States Informational and Educational Exchange Act" provided for "an information service to disseminate abroad information about the United States, its people, and policies promulgated by the Congress, the President, the Secretary of State and other responsible officials of Government having to do with matters affecting foreign affairs" (United States Informational and Educational Exchange Act of 1948, p. 6). Later, on August 1, 1953, the US government formed the United States Information Agency (USIA) out of the previous International Information Administration (IAA) formed in 1952. It was under the USIA that the "Voice of America" was formally named as one of five divisions within the agency (Senate Committee on Government Operations, Permanent Subcommittee on Investigations, 1954) and grew into a formidable operation in the post-war years with some of the world's most powerful transmitters and an extensive chain of relay stations broadcasting in 34 languages on nearly 100 frequencies ("Radio in the Cold War," 1954, p. 248).

Other radio broadcasting efforts in the war for hearts and minds following World War II included the project that came to be known as "Radio Free Europe" (RFE). Ostensibly the effort of private American citizens organized as the National Committee for a Free Europe, the service was launched on the Fourth of July in 1950 from studios at New York ("International: Anti-commie Programs," 1951) with full broadcast schedules commencing on Bastille Day (July 14). A periodical report of the time described the purposes of the service that transmitted from several European locations:

> Exiled democratic leaders of Europe will speak to their countrymen behind the Iron Curtain, freed of diplomatic restrictions and in their native languages. "They will give the lie to Communist propaganda and tell their listeners of the undying struggle to assure freedom everywhere," Dewitt C. Poole, NCFE president said. ("Radio Free Europe," 1950, p. 66)

Whatever the efforts at the time to portray the National Committee for a Free Europe, the founding agency for RFE, as an independent group of American citizens, most observers concluded that the hand of the US government was involved, even if not always evident despite the press using (conceivably tongue-in-cheek) references to NCFE as a "privately-managed organization" ("Czech Protest," 1951, p. 55) or to its "being privately backed and unrestricted by diplomatic protocol" ("International: Anti-commie Programs," 1951, p. 11). Declassified documents indicate that the founding of the NCFE was being discussed at the CIA at least as early as 1948 when the CIA's first director, Admiral Roscoe Hillenkoetter, held meetings that declared the agency's intention to work toward "the establishment of a democratic, philanthropic organization in New York under some such name as the American Committee for Free Europe which in turn would . . . organize a committee of responsible foreign language groups now in the western zones of Germany and provide them with facilities for communication with their homelands" (Hillen-koetter, 1948, p. 1). These would include funds, radio equipment and printing facilities to disseminate information that would include news, discussions of internal problems within target countries and "material designed to undermine support for the existing regimes" (Hillenkoetter, 1948, p. 1).

These discussions were, in part, a response to fears of the increasingly powerful role of radio in the post-WWII period, particularly in the hands of opponents such as the Soviet Empire. The United Press reported in 1947 that the Russians announced plans to construct 28 powerful new radio stations over the coming three years ("Radio Plans of Russia," 1947). The Soviet Union's Minister of Communication Konstantin Sergeichuk made the announcement on the occasion of Russia's Radio Day and noted that Russia was already broadcasting radio news and programming in 30 foreign languages and 70 languages used within the empire ("Radio Plans of Russia," 1947).

After being forced to admit its role in the NCFE and RFE (when it was exposed during the early 1970s) and to relinquish its control, the Central Intelligence Agency would openly tout RFE as being part of a covert government imperative in later years:

> On June 1, 1949, a group of prominent American businessmen, lawyers, and philanthropists—including Allen Dulles, who would become Director of Central Intelligence in 1953—launched the National Committee for a Free Europe (NCFE) at a press release in New York. Only a handful of people knew that NCFE was actually the public face of an innovative "psychological warfare" project undertaken by the Central Intelligence Agency (CIA). That operation—which soon gave rise to Radio Free Europe—would become one of the longest running and successful covert action campaigns ever mounted by the United States. (Central Intelligence Agency, 2007)

Less than a year after its launch, RFE expanded its operations to include a new station and broadcasts intended to "aim an anti-Communist campaign against Czechoslovakia . . . in Czech and Slovak language 11½ hours daily, in direct competition with the Communist-run Radio Prague and Radio Bratislava" ("International: Anti-commie Programs," 1951, p. 11). It was soon also announcing plans to expand its services to Soviet controlled Poland, Hungary, Romania and Bulgaria.

As might be expected, efforts such as those of RFE and the NCFE (which also included non-radio offensives such as the use of balloons that dropped propaganda leaflets into target territories) often met with resistance. While this resistance might take the form of signal jamming or counter-broadcasts, they also sometimes necessitated official complaints such as were reported in 1951 when "Czech authorities charged that US authorities had broken international agreements with 'hostile' broadcasts designed to foster espionage and terrorism" ("Czech Protest," 1951, p. 55).

The CIA-funded NCFE also undertook another similar project eventually called "Radio Liberty" (originally "Radio Liberation"), which, according to the CIA, "began broadcasts to the Soviet Union in 1953" and "rallied anti-Communist intellectuals, politicians, and activists to fight the Soviets in a contest for the peoples' minds and loyalties" (Central Intelligence Agency, 2007).

Eventually, external international broadcasts came to be seen as part of the United States' "national security and foreign policy goals" with the US government funding several programs including "the Voice of America (VOA), Office of Cuba Broadcasting (OCB), Radio Free Europe/Radio Liberty (RFE/RL), Radio Free Asia (RFA), and Middle East Broadcasting Networks (MBN)" (United States Broadcasting Board of Governors Special Committee on the Future of Shortwave Broadcasting, 2014, p. 2). Evidence of the enduring role of services such as the VOA, particularly in times of crisis or conflict, can be found in US President George Bush's request to Congress on September 10, 1990, for an additional $14.2 million in funding to support increased broadcasting, primarily through VOA, to the Middle East and specifically to boost VOA's broadcasts in the region from 7½ hours a day to 24 hours a day in both English and Arabic ("Bush Wants to Boost," 1990, p. 69).

However, VOA and all the US official external broadcasting services came in for review in the context of evolving media forms and usage patterns worldwide in 2014 when their controlling authority (an independent federal agency known as the Broadcasting Board of Governors) acknowledged the shrinking role of shortwave globally and the need for shifting to alternative media vehicles while maintaining the missions of these official US international broadcasters.

INTERNATIONAL BROADCASTERS IN CONTEXT

The evolution of international broadcasting was not limited to those highlighted here. There were many lesser efforts and some that were arguably more influential than these. The Soviet radio efforts, for example, were among the first to engage in deliberate broadcasts of persuasive messages to foreign lands (Graves, 1941, p. 13). The Soviets sought not only to promote cohesion over the Soviet Union's expansive land mass and to build its empire but also to spread its political messages to the rest of the globe and to argue its case to friends and enemies alike starting at the end of World War I. Wood (1992, p. 41) placed the formal start of Radio Moscow, "the external broadcasting service of the USSR," in 1929, "carrying propaganda broadcasts in German, English and French . . .

promoting the Communist way of life, telling its listeners about Soviet achievements."

These efforts largely failed to influence US audiences, not only because of a decline in shortwave listening in the United States, but also, according to at least one commentator, because the broadcasts committed the cardinal sin of being boring. Remenih (1951, p. A10) described Russian efforts over stations like Radio Moscow as "hitting us with everything in the book over Radio Moscow":

> Most Americans, lacking shortwave receivers, are unaware of the Reds' nightly propaganda blitz. . . . Typical of the vituperation the Reds are tossing our way currently was a commentary comparing President Truman to Hitler.

Three decades later, Russian broadcasts were still reaching the United States, with reports of AM band reception of "The North American Service of Radio Moscow" that was being relayed through Cuba in the early 1980s (Granville, 1981).

During the interwar period various currents of international influences also drove the exchange of news and information among nations, with the emerging technologies of broadcast radio being the primary means of immediate transmission. Even before the rise of dedicated systems of international broadcasting that would later be formalized by treaty and convention (that is to say, also including early commercial operations, amateurs and propaganda stations), international flows of news, information and political diatribes were common on the airwaves. Bent (1937, p. 117) pointed out that by 1937, the United States was communicating with some thirty foreign nations by various radio broadcasting channels and in doing so exchanged programming that, while nationalistic, fostered "better international understanding":

> During a single month recently we heard broadcasts both to and from China, Switzerland, Argentina, Australia, Austria, Bermuda, Brazil, England, France, Germany, Italy, Lithuania, Poland, and Czechoslovakia. King Haakon of Norway, King Christian of Denmark, King Gustav of Sweden, the Kings and Prime Minister of England, Premier Leon Blum of France, President Justo of Argentina, Adolf Hitler and Dr. H. H. Kung, acting executive head of the Chinese Government, have talked with us. In return President Roosevelt, members of his Cabinet, and United States Senators have been heard in alien lands.

While touting the idea that "it is through a better world acquaintance with individual national peculiarities of politics, law, personalities, and practices that an informed public opinion becomes possible," Bent (1937, pp. 117–18) also had to contend with the less than friendly exchanges evident in global radio of the time:

> From the Italian station at Bari, for example, anti-British broadcasts
> have been sent over a long period in Greek, Arabic, Serbian, Croatian,
> and Italian, as well as in English, to the countries of the Mediterranean
> littoral, such as Palestine, Egypt, and nearby countries in Europe, Asia,
> and Africa. Music by Verdi and propaganda for Mussolini take turns in
> programs for North and South America as well as for England . . .
> Poland, Jugoslavia [*sic*], Czechoslovakia, Rumania [*sic*], and Lithuania,
> the buffer states between Germany and Russia, are regarded as of great
> importance by the two larger countries, and are sprayed daily with
> propaganda.

Though the overt propaganda exchanges were major conduits of global
information flows (however biased), the seemingly benign exchange of
news and current affairs could also have their own impact since as Bent
(1937, p. 118) noted, "the great nations, including our own, chase each
other around the globe with their versions of news."

Cross-border broadcasts, both propagandistic and benign, also played
roles in the establishment of local stations where radio technologies
might not have otherwise been established. Sid-Ahmed (1984, p. 85) described the establishment of radio facilities in the Sudan as a response to
the flows of international broadcasts affecting the region:

> Radio Omdurman (*Huna Omdurman*) was founded on April 1, 1940.
> The British administration started radio transmission from a small stu-
> dio in the post office at Omdurman. . . . On January 3, 1938, the BBC
> started Arabic transmission to communicate with local citizens to
> counteract the Axis activity in Ethiopia and North Africa as part of the
> "Empire broadcasting." Radio Omdurman originated with a motive to
> serve the interests of the Allies and to defend against German-Italian
> propaganda and for dissemination of news and information.

Major Charles F. Atkinson (1931, p. 8), an official in various capacities
at the BBC over several years (including a tenure as its foreign liaison and
director of finance), wrote of European and international efforts to coor-
dinate broadcasting efforts during the early emergence of broadcasting in
Europe, partly due to the problems that emerged from signal interference
("listeners in each country were suffering atrociously from interferences
originated [quite innocently] in others") as broadcasters proliferated, but
also in order to address issues of propaganda:

> European broadcasting on a large scale began in 1923, and already by
> the following winter the situation was such that, a number of broad-
> casters spontaneously proposed common action. And thus the "*Union
> Internationale de Radiodiffusion*" came into being at a conference held in
> London in March, 1925.

Atkinson (1931, p. 8) also noted that, because of the international nature
of radio broadcasting, the union also grew to include numerous important non-European broadcasters as associate members. For all of this

breadth of membership, however, the union was thought to be relatively powerless and stood "accused of failing to prevent wireless propaganda" though some issues of cross-border propaganda were thought to have been "quietly settled within the Union under an internal gentleman's agreement" (Atkinson, 1931, p. 8).

The cross-border and global implications of international stations and others, as well as the surprisingly distant reach of early broadcasters, are important for many reasons. Not the least of these is the fact that they argue for a globalized media system that pre-dates both the digitally interconnected networks of the Internet and the globe-spanning reach of satellites. Even outside of the formalized system recognizable as international radio broadcasting and before some of the biggest names in that field were branded and established, radio was an international experience not just for those in the developed metropolises of the United States and Europe, but also for listeners in colonies and lesser-developed countries. As an example, Chua (2012, p. 170) described early listeners in Singapore having access to only "three amateur short wave stations run by individuals from the Amateur Wireless Society of Malaya" but still not being limited in their listening options:

> During this period radio listeners in Singapore were not limited to these local stations. Listeners were more likely instead to tune into literally an entire world of broadcasting via their short wave receivers. The daily radio schedules published in the newspapers show that in 1931, the Singapore listener's daily options included stations from Saigon in French Vietnam, and Bandoeng, Tanjong Priok, Batavia, Sourabaya, Medan, and Djokjakarta from the Dutch East Indies, as well as stations further afield in Melbourne, Sydney, Paris, Rome, Eindhoven, Zessen, Nairobi, New York City, and Moscow.

To be sure, there were distinctly different and powerful motivators of international radio broadcasting in the early to mid-twentieth century. The exchange of unvarnished propaganda in World War II and the competing strains of political discourse during the Cold War were only two of the most obvious drivers. The pressures to maintain far-flung empires against external threats and internal dissent were also important motivators to European colonists, while in a parallel sense, the need to keep contact with diaspora populations outside of India was important to AIR.

Broadcasters such as the BBC World Service and the Voice of America even collaborated to protect their global interests, resulting in shortwave services that addressed audiences in regions perceived as being at risk for hostile influences. In 1946, *The Wall Street Journal* reported on an interactive broadcast prompted by global forces and featuring cooperation between "Uncle Sam and John Bull" who "since December 1, 1942, were on the air with a most unusual program" beamed to the Caribbean:

> The half-hour program, highly informal, is called *"The West Indian Radio Newspaper."* Its sponsor is the Anglo-American Caribbean Commission, an intergovernmental agency. From Washington the entertainment is piped by wire through New York to Boston where it is shot into the ether by two transmitters, WRUL and WRUW, whose direction antennae blanket the Caribbean. (Estill, 1946, p. 1)

Programming on this multinational broadcast was always subject to change due to war conditions but included titles such as *The West Indian Radio Newspaper Symphony Concert, Letters from Listeners, Creole Cook, Agricultural Chat, Americana, Popular Concert* and *Caribbean News and Music* (Boord, 1944, p. 54).

International radio broadcasters have generally occupied a space of some ambivalence. In some places, they have been perceived as tools of propaganda and foreign influence. It is, and has been, difficult to prevent these broadcasts from crossing, even when unwanted, over vast distances and national borders. In many cases, they would fill information vacuums for audiences who lived with scarce information sources or restricted access to information (Fiedler & Frère, 2016). Hale (1975, p. x) has argued that where censorship has limited access to information, international broadcasters have served as "the only source of a large part of people's information about things that happen outside their immediate locality and their closed society."

International broadcasters are also important in times of conflict or disaster when local communications are unreliable or unavailable. A simple example of such roles can be found in the broadcasts of the Voice of America during the first Gulf War in which VOA broadcast warnings to Americans and others living in the vicinity of Kuwait and Iraq against trying to flee by crossing the desert areas unassisted and providing information on sources of assistance and evacuations.

Where their signals crossed into authoritarian regimes or competing ideological spaces, the programming of international radio broadcasters, often made accessible through multiple language services, often encountered politicians and generals who decried the content as disinformation or propaganda (as they often were). When these broadcasts provided information contrary to government or official sources, authorities might even deem them to be threats to national security while often depending on them for independent information from outside.

The case of cross-border broadcasts from the United States to Haiti demonstrated some of these dynamics (including language accessibility) in an instance where the US government was not the instigating party. In September 1965, Haitian dictator François "Papa Doc" Duvalier complained about shortwave radio broadcasts emanating from Radio New York Worldwide (WRUL Shortwave) in New York City and beamed to Haiti starting in July of that year, characterizing them as "aggression against his government":

The announcers speak in Creole, the Haitian patois, under sponsorship of the United *Haitienne Internationale,* an exile group headed by Paul Magloire. Magloire is a former president of Haiti who was ousted in 1956, the year before Duvalier came to power. The controlled Haitian press has not mentioned the broadcasts or Duvalier's protest note. Nevertheless, many Haitians listen and the program is a lively topic of conversation. . . . Use of the Creole language is a significant concession to the Haitian peasants. Most of them cannot read Haitian newspapers or understand Haitian radio station announcers because they use French, the official national language. Only educated Haitians speak French. ("Haiti's Chief," 1956, p. D10)

The newspaper report went on to note that the Haitian peasantry also depended on Haitian exile radio from Cuba for their information about Haiti and the outside world since those broadcasts were in Haitian Creole as well.

By far the most prominent of these narrow radio efforts directed specifically at a small population and emanating from a powerful neighbor was *Radio Martí,* which was clearly a government-backed effort from US authorities (under the authority of the Broadcasting Board of Governors) and clearly intended to influence Cuban audiences against the Castro regime. The station started transmissions in May 1985 and was perennially condemned as an outdated propaganda machine but simultaneously praised as an expression of solidarity with Cuban dissidents. Also, by the time the Reagan administration announced plans for the station in 1981 with plans for budget appropriations approximating US$10 million for 1982 and US$7 million for 1983 (United States Senate Committee on Foreign Relations, 1983), both government and private interests in the United States had already established some two decades of anti-Castro radio broadcasting efforts in various forms (Nichols, 1984).

Outside of these very specific and directed radio efforts, the broader flow of news and information remained the domain of the major international radio broadcasters. Several of the major international broadcasters, particularly in postcolonial nations, found themselves to be reviled and revered at the same time. As sources of reliable information, broadcasters such as the BBC World Service and the Voice of America were held in the highest esteem with their news and programming routinely finding either direct audiences or being relayed as part of local media news. Boyd (1999a, p. 5) noted that even as late as the 1990s, these international broadcasters enjoyed a reputation for reliable information accessible in numerous languages, particularly where local and regional media were known to demonstrate biases and particularly during times of crisis:

People in the Arab world can fairly reliably receive, even on inexpensive transistor radios, programs from the Voice of America, the British Broadcasting Corporation, and Radio Monte Carlo Middle East (France). Surveys indicate that these and other services are avidly lis-

tened to in the Arab world by followers of Arab and international
events. . . . Listeners understand that the government has its priorities
and its points of view, and those who listen to non-Arab international
broadcasts appear to be listening in order to gain another opinion
about a local or international event. Listening during times of crisis is
an almost standard procedure in the Middle East, particularly among
the elite.

Fandy (2007) noted one such event that accounted for an enduring
lack of trust in domestic and regional media followed broadcasts in 1967
by Egyptian state radio (echoed by numerous Arab government-owned
media operations throughout the region) indicating that the Israeli army
had been defeated and that the Arab forces had prevailed in the Six-day
War. The truth of the Arab defeat was learned from foreign radio broadcasters such as the BBC.

In the role of external providers (sometimes of internal news), these
international radio broadcasters, while certainly not seen as unbiased,
were still useful sources of information that could be expected to
transcend local and regional biases and provide some level of factual
account. They served as important sources of information on major international events that would be inaccessible, expensive or politically sensitive for local media to cover. Yet, these media sources also represented,
for former colonies, a continuing dependence on the former colonial masters for a resource (information) and the continuation of colonial systems
of authority in which the word of the colonial master held sway over the
discourse of the locals.

As former colonies engaged with the various emerging questions of
nationhood, it became evident that many of these so-called "international
radio broadcasters" also tended to reflect a decidedly metropolitan and
European view of the world, ignoring issues and areas important to so
many in their global audiences. The irony of distant listening for regional
news was not lost on these former colonies, and so the fact that Radio
France International, for example, was often the intermediary by which
North African nations might hear news from their regional neighbors,
caused increasing concerns in many former colonies and prompted the
development of a spate of alternatives to international broadcasters, partly embodied in regional news agencies and broadcast exchanges such as
the Caribbean News Agency (CANA) and the Pan-African News Agency
(PANA).

The case of All India Radio here presents a relatively special example
of the impact of such geopolitical change on traditional public global
broadcasting. In India, international broadcasting, originally developed
partly as an imperial counter to foreign propaganda, survived into that
country's independence and evolved into somewhat different roles there-
after. Issues of Indian identity, for example, brought into question whether content such as Hindi film music represented appropriate program-

ming for their global voice. Ironically, the decision that such commercial content was not Indian enough was probably responsible for a loss of many listeners in their diaspora who tuned in for just that fare.

Though the established names such as the VOA and the BBC World Service dominated the field of international shortwave broadcasting, it was not solely their domain. A few private commercial attempts at international shortwave broadcasting also emerged during the post-war years. The US network CBS, for example, had established itself as a major international shortwave broadcaster, setting up a shortwave bureau in 1937. CBS also ran a Latin American service that ran from 1942 to 1948 and boasted as many as 126 stations ("CBS Radio," 1968). The US government restricted these facilities from independent broadcasting during wartime (when they were permitted to relay material from the Office of War Intelligence) and, after the war, purchased most shortwave facilities for the Voice of America (Schneider, 2014). In 1968, CBS announced that it would again beam programming to Latin America on shortwave using different facilities. However, no such commercial efforts at shortwave radio were ever commercially successful.

Technical challenges to shortwave also emerged. Even in peacetime, conflict over broadcast signals was inevitable. International broadcasters often had to hold international coordinating conferences to negotiate signal strengths, directions, relays and other technical details to avoid interference that could cancel out one another's signals to a particular audience. Added to these difficulties, various nations began to invest in local medium-wave broadcasting and even FM stations that brought high-quality reception to audiences on a local scale. With such developments the average listener no longer had to cope with the prospects of poor reception and poor audio quality, while the availability and allure of local content began to erode away at the need for shortwave signals.

Neither the changed geopolitics (including the end of World War II and the Cold War) nor the changed technological landscape have necessarily spelled the end of the state-run international shortwave broadcasters. For many of them, changes in geopolitics and technology have simply meant that the broadcasters have had to adapt in order to maintain their missions of spreading the particular view of their nation to others. During the 1990s, for example, several of these stations, traditionally found on shortwave bands, began to migrate to the World Wide Web. Many have diversified their radio operations into a more multimedia web presence to compete with emerging media forms.

It is also worth mentioning that while major names such as the BBC and the Voice of America often demand greater historical attention, numerous other smaller operators were, at times, involved in international broadcasting, often in deliberate contravention of licensing or other restrictions. One of the earliest of these so-called "pirate" stations emerged in May 1933 as station RXKR began to broadcast from a Panama-licensed floating casino offshore toward Southern California until the US Coast Guard shut them down in August (Walker, 2001). In 1964, perhaps the most famous pirate station, Radio Caroline was launched from a ship off the British coast, in part challenging the BBC monopoly on broadcasting and BBC claims that there was no demand for commercial radio in England. More often than not, these pirate operations have shown more interest in overcoming local restrictions on broadcasting rather than aiming at global reach.

FIVE
Local Radio Broadcasters and Their International Audiences

The global flows of broadcasts and influences, particularly during the early experimental stages of shortwave, produced numerous examples of unlikely content and signal exchanges. One such example involved the global connections among Holland and its colonies and how those would intermingle with the global connections among Britain and her colonies. Berg (2013, p. 50) recounted the earliest Dutch experiments:

> Private experimental station PCGG operated from 1919 to 1924. . . . It had many listeners outside Holland, in particular the UK [The *Daily Mail* sponsored an English-language concert series over the station]. . . . In 1926, researchers at *N. V. Philips Gloeilampenfabriek,* a lightbulb maker turned major radio parts manufacturer and reseller, located in Eindhoven, began shortwave broadcasting experiments with a 300-watt transmitter on 3313 kc. PCJJ as it was called, was heard throughout Europe. . . . PCJJ's goal was to provide a regular shortwave service directly to listeners in the Dutch East Indies, and to be heard as well in the Dutch colonies of the western hemisphere, the Dutch West Indies (the Netherlands Antilles) and Dutch Guiana (Surinam).

Modern Wireless magazine reported that the station, soon after its launch on April 28, 1927, featured addresses from important Dutch officials and leaders to their colonial audiences:

> On May 14th, 1927, the Minister of the Colonies addressed the Dutch Colonies through the Philips transmitter. On May 31st and June 1st, 1927, Her Majesty the Queen and Her Royal Highness Princess Juliana visited Philips Short-Wave Station and, via the microphone, they addressed their subjects in the Dutch East and West Indies. ("A Notable Broadcast," 1928, p. 156)

The unlikely dimension of this broadcasting effort was not that the Dutch aimed their signals at their colonies such as Suriname, but the fact that these broadcasts would figure prominently in the emergence of radio in Suriname's neighbor, the British colony known at that time as British Guiana.

GUYANA

The present discussion has taken British Guiana (now Guyana) into account above, in part due to its early adoption of radio technologies in varying forms. Whereas most histories of radio and explorations of the global scope of radio tend to focus on the roles of broadcasters from European nations, important developments in radio were also taking place in other places. We have mentioned India, for example, which developed an international broadcasting system out of its work to reach its own population. Yet even those places that did not evolve into formalized international broadcasting did also experience the cross-border reach of radio and its ability to unify distant audiences. Guyana was one such location whose experience with radio as both a broadcaster and a receiving audience demonstrates radio's global power from its early days to bridge distance and engage diverse listeners in distant places. It will be necessary in the discussion going forward to switch between the terminologies of British Guiana or Guiana in historical references and Guyana in modern sources.

Guyana is located in northeastern South America. Its capital, Georgetown, lies about 4,500 miles (7,241 km) from London, England, 2,500 miles (4,000 km) from New York City and 4,700 miles (7,600 km) from Amsterdam. Located, as it is, on a large landmass, the territory is quite large compared to other British colonies in the region, many of which were small islands in the Caribbean Sea. Bearing in mind that its neighbor Venezuela lays claim to substantial parts of Guyana's jungle areas, the official land area of Guyana is roughly 83,000 square miles (215 km^2). Guyana's history includes influences from its native peoples who have lived there since before recorded history, Dutch colonial occupation and British colonial occupation which brought several groups of people to the territory including Africans as slave labor and, later, Indians as indentured laborers for the sugarcane plantations. Smaller groups from other global regions also found their way to Guyana including Portuguese and Chinese laborers with whom the British experimented as plantation laborers. British Guiana gained independence (as "Guyana") from Great Britain in 1966.

The emergence of early radio in British Guiana happened under British colonial rule in the period between World War I and World War II. British authorities equipped British Guiana with wireless telegraph equipment as part of their efforts to improve their wireless telegraphy and maritime communications capabilities around and after World War I.

The potentials of wireless telegraphy expanded well beyond the coast, however, as this distance-reducing technology proved to be a useful component of Guiana's mining operations, which include lucrative gold and diamond mines. During the 1920s, miners began to employ wireless communication systems to connect base operations to mining sites deep in the jungle interior. For this reason, wireless and radio were included in the legal frameworks surrounding establishment and operation of the mines. The US Bureau of Mines information circular on the mining laws of British Guiana, for example, noted (Youngman, 1934, p. 14) that "The Governor may grant to others than the concessionaire, lessee, or licensee the right to construct a pumping plant, hydroelectric plant, pipe line, transmission line, telegraph line, wireless or radio station." A newspaper report from January 1927 recounted that a disruption in the wireless communications system was interfering with the smooth operation of transportation and mining operations ("Men Going," 1927).

Apart from point-to-point, two-way radio communications, the notion of broadcast radio was gaining popularity in the early 1920s. The residents of British Guiana at that time included British colonials (some visitors, others who were born in the colony) and the descendants of African slaves and Indian indentured laborers. The British controlled all major aspects of the colony including its government, economy and judicial system. Wealth was concentrated in the hands of the British colonial masters, and the initial impetus for radio reception was restricted to the wealthy and privileged colonial upper crust, including those who lived in the capital of Georgetown and were able to afford both telephone lines and the luxury of a broadcast relay subscription.

Residents of the colony of British Guiana, though in a remote corner of the British Empire, were not cut off from information about the rest of the world. By 1893 the colony boasted both an undersea telegraph cable connection as well as 260 miles of telegraph lines (Greswell, 1893, p. 270). The colony boasted wired and wireless telegraphy (a station having been established by the British Navy in 1909) as well as a telephone service (starting in 1884), regular ship traffic and local newspapers (established in 1880 and 1881) capable of receiving international news by cable and by wireless.

All these modes of communication ensured that Guianese were in tune with developments abroad, including the development of radio. Among enthusiasts and experimenters, the possibility of receiving signals, however distant, was an emerging challenge in the 1920s. Yet, the reality was that signals were relatively scarce and weak and radio was not yet a dominant form of information or entertainment. Predominant forms of entertainment prior and up to 1927 in Guyana included movies at theaters such as the Gaiety in Georgetown (Chin, 2009), public (including open-air) band concerts and traveling circuses.

Typical "Band Programmes" of the time included British Guiana Militia Band afternoon performances at the Sea Wall in Georgetown including selections such as Weber-Weingarten's *Invitation to the Waltz* and Michaelis' *The Turkish Patrol* (Fawcett, 1927). However, as these selections indicate, there was a decidedly European and colonial bent to these performances. As Cambridge (2015) has aptly noted, these performances from the British Guiana Militia Band at locations such as the Sea Wall, the Botanical Gardens and the Promenade Gardens in Georgetown catered to the ruling European elites in the colony.

Somewhat less stodgy were occasional public entertainment displays including a variety of attractions such as an advertised "Patriotic Garden Party" in January of 1927 at the Promenade Gardens that was carded to feature "Gouvia's popular jazz band," "instrumental and vocal songs," "acrobatic feats by Lustal Henson Company," "fancy costume parades to music" as well as "attractive side shows and entertainment" all with "brilliant illumination" ("Patriotic Garden Party," 1927, p. 8). Admission for such an extravaganza was one "bob" (shilling). *The New Daily Chronicle* of February 4, 1927, reported on the arrival of a traveling circus to less than positive responses in Georgetown:

> One of the queerest comedy troupes which have ever been seen in this City arrived in Georgetown yesterday shortly after 1 p.m. It speaks volumes for the remarkable development of circus Shows in relation to economy and the lightening of a once cumbersome train of actors that Hon. R. E. Brassington, proprietor of the Show, was able to stage a few minutes after arrival here a performance, in the streets of the city, of the well-known and exacting burlesque "Sly Mongoose." The great Showman travels only with a young monkey, two motor cars, and a band of *Kallimai Pooje* music. ("Comedy Troupe," p. 4)

In 1927, the challenges of reception combined with excitement over the emerging availability of broadcasts both in Great Britain and in other parts of the world prompted the creation of a wired system of broadcast relay available to telephone subscribers in the capital, Georgetown ("Local Broadcasting," 1927; Sanders, 1978). Under this system government facilities with powerful antennas received foreign broadcasts and relayed them via phone lines to subscribers who received the relayed broadcasts on connected, purpose-designed boxes in their homes. Most of these subscribers were wealthy expatriates who could afford both a telephone service and the extra service fees.

The telephone relay system was in place at the time of the 1928 world heavyweight championship boxing match between Thomas Heeney of New Zealand and defending champion James Joseph "Gene" Tunney of the United States in New York City on July 26th. Local newspapers in

Guyana reported that the Georgetown telephone relay would carry radio broadcast coverage of the fight live from New York, but with some competition from experimental broadcast facilities that were in place ("Local Broadcasting Service," 1928, p. 7). Newspaper reports also revealed something of how users might experience the system and the actions required to secure a listening channel:

> The fight in Madison Square Garden for the World's Heavyweight Championship will be relayed on Thursday evening over the Georgetown Telephone Service and simultaneously by means of the experimental broadcast transmitter. . . . Regular subscribers to the Telephone Relaying Service should notify Central of their intention to listen to the fight as early as possible, in order that arrangements may be made to cater for the correct number. ("Local Broadcasting Service," 1928, p. 7)

This system was short-lived, failing to achieve either financial or technical success and lasting for a little over one year (Sanders, 1978). However, its accomplishment and its role in stimulating further interest in radio broadcast reception are relevant to the development of radio in Guyana. Indeed, the wired system started the notion of radio schedules and popularized the notion of listening to broadcasts. Published schedules in local newspapers of the time included program listings for KDKA in Pittsburgh, as well as stations in Atlantic City, New Jersey, Schenectady, New York, and the Eindhoven station from the Netherlands run by the Phillips Company, which had added English programming to its transmissions.

The fact that the wired subscribers were listening in to broadcasts from Eindhoven was clearly the result of Dutch efforts to broadcast to their own nearby colony Dutch Guiana (Suriname). Additionally, even after the BBC began its regular international broadcasts specifically aimed at its own colonies, the local newspaper continued, for several years, to provide listings for external stations including those at Schenectady and Pittsburgh.

Broadcast radio would take a few more years to evolve in Guyana with a few false starts along the way. In June of 1928, local newspaper reports indicated that the local telecommunications officials had set up an experimental broadcast station with the call letters VRY ("Local Broadcast Experiments," 1928). This facility primarily broadcast the only content then available, which was the existing (though hard to receive without large antennas and powerful tuners) foreign broadcasts. This was the same experimental transmitter involved in broadcasting the Heeney-Tunney fight in July 1928 and local officials sought listener feedback on the reception at the time as they continued to program through the weekend following the fight ("Local Broadcasting Service," 1928, p. 7):

> Tests of the experimental transmitter will be continued during the week-end, and a Band Program will be relayed from the Sea Wall on Saturday afternoon from 5 to 6 o'clock. The third Sunday evening test

concert will also be given Sunday evening, commencing at 8 o-clock; listeners will be very materially helping the progress of these experiments by reporting conditions of reception to the Post Office Engineering Branch.

Though sporadic in nature, the broadcast experiments continued for some time until they abruptly stopped in 1931 due to a lack of funding. Amateurs and hobbyists, however, continued to experiment with broadcasting and kept the local desire for radio alive.

The desire for local broadcasting forced more pragmatic measures to produce the first attempts at a commercial station. This happened only after tensions between the colonial authorities' hesitation to establish a substantial broadcasting presence and public desire for such a facility came to a head in 1935 with the erstwhile hobbyists and amateurs coming into their own and becoming, *de facto*, professional broadcasters.

When the celebrated British cricketing team known as the Marylebone Cricket Club announced plans to tour the West Indies and British Guiana, local fans began to clamor for the ability to follow this tour using radio. Angry letters and articles in the press bemoaned the fact that other colonies such as Trinidad and even little Barbados were beginning to show signs of developing broadcast capabilities while British Guiana sadly lagged. Trinidad's test broadcasts in January of 1935 caught the attention of the British Guiana newspapers with a report datelined January 12th that noted:

> Another successful trial broadcast was again made on January 11 when Mr. Diego Serrao, the local amateur, was on the air between 10 o'clock and 11 a.m. Reports received showed that not only was the Trinidad broadcast from Station VP4Y6 heard throughout the City, but also in St. Vincent, St. Kitts and in British Guiana, though reception in Demerara was reported to be not of the best. ("Trinidad Broadcast Test," 1935, p. 8)

Not a day too soon, however, on February 4, 1935, on the eve of the opening match of the series, government officials (including His Excellency the Officer Administering the Government, Sir Crawford Douglas-Jones, CMG) proudly gathered (with members of the visiting cricket team) at a private photo studio (also dubbed the Crystals Broadcasting Studio) on Wellington Street in Georgetown to congratulate private amateur operators for bringing to the air the amateur station VP3BG ("Station 'V.P.3.B.G.,'" 1935). His Excellency used the opportunity to welcome the visitors and to emphasize the role of cricket in establishing the broadcast facility:

> I wish all success to the Amateur Radio Station "VP3BG" and congratulate Mr. La Motte Kerr and those associated with him in the enterprise they have shown in establishing this station. . . . The visit of the M.C.C. Cricket Team to the West Indies and British Guiana has created a de-

mand for broadcast comments on the matches during their progress. The comments broadcast from neighboring colonies have been eagerly listened to. ("Amateur Broadcast Station," 1935, p. 5)

When the visiting British Marylebone Cricket Club faced off against the British Guiana XI at the historic Bourda cricket ground in Georgetown on February 5, 1935, the local press indicated that the match would be carried on local radio with running commentary and reviews of play ("The Opening Match," 1935). Soon after the launch of VP3BG, another station run by amateur interests turned media entrepreneurs came on the air as VP3MR with the slogan "The Voice of Guiana" and an address of Luckie's Chambers, Georgetown ("Station List," 1937). Sanders (1978, p. 21) characterized the evolution of the two early British Guiana stations stemming from the coverage of the MCC cricket matches, writing that:

> Two stations, VP3MR and VP3BG were operated independently from this time on a commercial basis with sponsored programs until 1938 when they were amalgamated on the formation of the BG United Broadcasting Company Ltd which was financed by local firms and individuals. In 1949 a medium-wave transmitter was brought into service in addition to the shortwave transmitter.

By 1937, VP3MR, broadcasting as "The British Guiana Broadcasting Company, Limited," was on the air from Monday to Saturday from 5:30 to 9:30 p.m. and on Sundays from 9:00 a.m. to 12:00 p.m. local time ("Impressive Veri," 1938). The *New York Times* of January 17, 1937, mentioned the two Guyanese stations among those radio signals being received in the continental United States on the shortwave band:

> Station VP3MR at Georgetown, British Guiana, is heard clearly these days on its new frequency of 5.98 megacycles. With the exception of VP3BG on 6.135 megacycles, VP3MR is the only English-speaking short-wave broadcaster on the continent of South America. ("Radio's Short Waves," p. 12X)

In 1938, these two stations merged to form the British Guiana United Broadcasting Company Ltd. operating as station ZFY, Radio Demerara out of Georgetown with offices eventually established at the corner of High and Princes Streets in 1955 (Radio Demerara, 1960). ZFY became much more than a relay service for international shortwave broadcasters. Press coverage and station documentation reveal that the station carried local live events including public concerts and church services as well as studio productions such as quiz shows in addition to news and music. In 1945, the station programmed with several types of material reflecting its domestic and international dimensions. These included local productions, BBC rebroadcasts and transcriptions, US Office of War Information and US Armed Forces programming as well as rebroadcasts of the previously mentioned West Indian Radio Newspaper. Its colonial status was also evident in the fact that all its transmissions opened with *"Rule Britannia"* and closed with *"God Save the King"* (Boord, 1945).

From its inception, ZFY was more than just a local station and its regional reach and importance (though perhaps not sufficiently emphasized in the historical literature) has, at times, come in for mention. In 1945, for example, station manager G. V. de Freitas described British Guiana as ZFY's primary territory but noted that its secondary coverage target was the British West Indies from Antigua to Trinidad (Boord, 1945). The islands of Trinidad and Tobago (both forming another British colony at the time) lie about 350 miles (579 km) to the northwest of Georgetown, Guyana. These islands experienced a similar influx of Indian indentured laborers (often called "East Indians" in the colonies) during the 1800s and early 1900s under the same scheme that brought this group to British Guiana. This fact accounts, in part, for what *The Guyana Review* has noted (in keeping with several online memoirs and other accounts) as the regional reach and popularity of British Guiana's ZFY:

> ZFY reportedly had a significant Trinidadian audience being for them the main or only source of religious broadcasting and of Indian musical entertainment. Its popularity with Trinidadians, it seems, persisted even after September 1947, when Radio Trinidad was inaugurated. . . . In 1951 the station became Radio Demerara. ("A Brief History," 2010, p. 14)

Numerous older persons of Indian ancestry from Trinidad and some radio pioneers confirmed the fact that they and their families did listen to radio from "B.G." for Indian music. Most of these respondents were not old enough to remember pre-1947 so that they referenced the post–Radio Trinidad period. However, more concrete evidence of the regional scope of Guyanese radio does exist in the form of a little-mentioned collaboration between radio pioneers in British Guiana and Trinidad.

The story of this collaboration begins with a man named Mohamad Akbar who was born in 1889 in Stanleytown just outside of Georgetown in British Guiana. Akbar would become involved with an organization called the British Guiana East India Association (formed in 1916) as the first editor of that group's newspaper. With a background in matters of the diasporic East Indian culture and some acumen for business, Akbar went on to become established as a radio personality with ZFY, specializing in Indian music. According to his letterhead, Akbar established a program entitled *The Indian Hour of National Music* in 1937 (Akbar, 1946) and was described as the "advertising agent and sponsor" of the program on ZFY (Daily Chronicle, 1948). One reason for this description was that the hosts of such programs routinely acted as advertiser by both soliciting sponsorship from businesses and also by charging listeners to play dedications on the air.

Letters from Akbar appear in the papers of Trinidad's own radio pioneer of Indian music broadcasting named Kamaluddin Mohammed (born in 1927 in El Socorro, Trinidad). These correspondences indicate that not only was the Trinidadian listening to Mohammad Akbar's radio programs from British Guiana but also that Kamaluddin harbored aspirations to start broadcasting his own program when radio came to Trinidad. When the opportunity to do so came after the launch of Trinidad's radio station in 1947, Kamaluddin sought and received feedback from Akbar on the quality of his program and on the quality of the signal reception in British Guiana (Akbar, 1947). The two would continue to collaborate, forming a regional, cross-colony partnership that involved visits to the other territories and programming with distant audiences in mind (Akbar, 1947; Khan, 1947).

Radio Demerara opened new studios with (then) modern equipment in 1953 and boasted just prior to the colony gaining independence of the fact that "one of the most pleasing features in the progress we have made is the fact that Radio Demerara employs 99 percent Guianese and West Indian people" (Radio Demerara, 1960, p. 3). This sense of regional and national identity implicitly framed against the looming history of colonial dominance in all facets of social development would be manifest throughout postcolonial societies and in their evolving media. Guyana was no exception.

Media, Politics and Postcolonial Development in Guyana

Many independent nations emerged out of the colonial domination of European powers in the years following World War II. Several conceptual approaches popular from that time forward shared a general notion of mass media as a tool of national development from the 1960s onward. These included the development communication school under writers such as Schramm (1964) and Rogers (1976) as well as more strident and political voices such as those of Walter Rodney (1973), Manley (1987) and others in the so-called "dependency" school which also viewed foreign media as an inherent threat to identity and economy. A more general and mainstream approach that viewed mass media along with literature and other forms of expression (as well as politics and social assumptions) through the lens of the colonial past and independent futures has been termed "postcolonial theory" and is associated with writers such as Said (1978) and Fanon (1961, 2004) (though the latter is also highly regarded among dependency theorists). Postcolonial theory has to do with the persistent effects of imperialism particularly as embodied in global institutions, relationships of power and the persistent economic paradigm in which the colony continues to be treated as a tributary for the traditional colonizer (Beckford, 1972). While these currents of thought swirled, the emerging policies of developing nations with regard to their investments in and use of mass media struggled to cope with competing demands that they be at once involved with and insulated from global centers of power. Much of their decision-making with regard to their own radio and with regard to content from outside was perceived through the lens of postcolonial thought.

Great Britain, under severe economic strain and the pressures of resistance to its rule in places such as India, began a process of divesting its colonial control throughout its vast empire. With spectacular failures such as its partition of India, the British attempted to proceed with some caution in other areas and gradually introduced systems of colonial self-rule in other colonies such as those in the West Indies and South America during the 1950s.These experiments were not always successful and often had unintended consequences, exposing, for example, racial strife in places such as British Guiana and Trinidad where populations eventually coalesced around parties defined more by race than political ideals.

Premdas (1971, pp. 26–27) described this postcolonial political evolution in Guyana emerging out of an "original nationalist movement" that was "spearheaded by a mass-based, multi-racial party, the People's Progressive Party (PPP), formed in 1950 and led by Cheddi Jagan and Forbes Burnham":

> The PPP succeeded in setting in motion the independence struggle, but before it could achieve its objective, the two communal leaders, Jagan and Burnham who represented the Indian and African communities respectively, quarreled and parted company over differences regarding strategy and tactics of the movement . . . each leader established his own party which persists to the present. But each party is fundamentally different from the original multiracial PPP. Jagan's new party, the PPP (using the name of the original party) and Burnham's new party, the People's National Congress (PNC), represent almost exclusive solid racial blocs.

Another dimension of insularity also prevailed regionally when attempts to forge a regional federation of states consisting of former British colonies in the West Indies failed. Under this proposal, the former colonies would be granted their independence as a political union of states rather than individually. However, internal strife over such issues as the location of the federal capital and concerns about how power was to be distributed led to the dissolution of this plan by 1962, after which individual territories each pursued their own independence.

As all the territories of the region debated the prospects of emerging into independent nationhood, serious objections arose over the continuation of colonial dominance. This postcolonial resistance manifested itself in different ways (as it did in many different nations around the world), including various strains of nationalisms and the embrace of new ideas and ideals. The new national state, divested of both the empire's domination and its benefits, was a source of both hope and trepidation. Fundamental questions about relationships such as those among government, industries, unions and workers were (while somewhat prescribed by colonial tradition and institutionally biased toward perpetuation), to a great degree, up for renegotiation.

For British Guiana, strong Marxist and socialist currents began to swirl early in its colonial politics (even prior to independence), a situation that was less than comforting to the British authorities and of more than a little alarm to its allies, including the United States. These currents of Marxism and socialism met the postcolonial uncertainties of local politics and a backlash against all vestiges of colonial domination (including the somewhat glaring history of slavery) expressed in what came to be known as the Black Power movement. This movement came to greatest notice in the region during the late 1960s and 1970s with the work of outspoken and controversial Guyanese advocate Walter Rodney. While this was not the only political and social force in the region, the Black Power movement was a potent (and, rarely, violent) response to persistent colonial systems and attitudes related in no small part to racial attitudes, privilege and power (Lowenthal, 1972) made more complicated in the newly independent Guyana by its internal composition of Indo-Guyanese and Afro-Guyanese sub-populations.

Added to these potent global forces was a somewhat more bureaucratic, but equally oppositional movement involving the redefinition of the roles and functions of mass media in a world that increasingly comprised independent nations rather than empires and their satellites. With the increasing development of electronic media, particularly those with cross-border and international reach, new, emerging and lesser-developed nations began to express their concerns over the fact that these electronic media were dominated by the same powers who previously dominated their countries. Former colonies and other developing nations saw the continuing dominance of these foreign powers in electronic media and international news systems as a threat to their own developing independence as these major international news media dominated ideas about the world, portrayed developing nations negatively and perpetrated one-way flows of information.

Additionally, the fact that former colonial powers and global superpowers dominated so much international media policy and infrastructural decisions created a perception among developing nations that they were being forced to abide by rules that did not serve their own interests. In these former colonies and in developing nations more broadly, the emerging electronic media (that at the time went beyond just radio and included television, often aided by satellite links) held the potential to be used as tools of national development. Foreign academics including mass media stalwarts like Wilbur Schramm (1964) fueled this perception with works outlining how these media had been and could be used for social development, particularly in overcoming the resource limitations in poor countries. Soon, however, the notion of mass media for national development would become a matter of contention at international meetings, and the United Nations Educational, Scientific and Cultural Organization (UNESCO) was the epicenter of these battles where developing nations fought for what was called a "New World Information and Communication Order."

While some scholars argued that the continuing relations of power in mass media systems amounted to a kind of cultural imperialism (Schiller, 1976) or electronic colonialism (McPhail, 1987), leaders of many developing countries saw the insistence of former colonial powers and global superpowers like the United States on freedom of the press as a foreign cultural imposition. Zaffiro (1984, p. 49), describing this evolution in Africa, wrote:

> In political terms, radio became revolutionary in Africa when receivers were no longer pre-set to the colonial station, beginning in the 1950's. To the right or left on the frequency band a score of stations could be reached. As a medium of political communication, radio offered a means of comparing information; radio served as a daily reality test. For some, owning a radio and listening to external news broadcasts became political acts, revealing a desire to take in other prospects and points of view.

Similarly, Domatob and Hall (1983, p. 13) described:

> A common pattern in most African countries consists of a central government Ministry of Information with radio and television senior personnel being civil servants who report to the Minister of Information. . . . Radio broadcasting is of course the primary medium of mass communication and, in Africa as elsewhere in the Third World, is generally controlled by the regime in power.

In the specific case of Cameroon, Domatob (1985, p. 124) wrote that the colonial radio infrastructure set in place since 1939 took on a development role at independence in 1960, viewed as a "legitimate tool to pursue national goals":

> The first objective was the survival of Cameroun as a single entity. . . . With the media's aid, political leaders focused attention on rousing Camerounians to the new cultural imagery of nationhood reminiscent of the goals of nationalist movements in 19th century Europe. Cameroun had a President, flag and national anthem, the problem was how to instill loyalty among the scores of cultures, ethnic and linguistic groupings scattered across the nation. Getting the message of national unity across was the media's job.

These and many other developing nations responded to the perceived imbalances in news and cultural flows with moves to bring the media, perceived as important national assets, under government control to be used in the interest of the nation as a whole and in the creation of resistance to the foreign domination implicit in mass media controlled by the already dominant forces of former colonial rulers and international capital.

In newly independent Guyana (as in Trinidad and Tobago and other regional states), governments increasingly touted greater national control of resources in general as a natural evolution of their independence. Among the resources to be nationalized, they often included the media, with its potentials for national mobilization and to counter external influences. In the case of Guyana, its particular set of pre-existing political ideas gave less credence to the notion of a free press being important for democracy and more focus on the importance of the media as a national resource. In already socialist-leaning Guyana, this nationalist turn was taken almost for granted. As Guyana embraced its transformation into what it called a "Cooperative Republic" (which it declared itself in 1970), the role of mass media for national development was all but entrenched; a 1979 front page of the *Guyana Chronicle*, for example, announced that "Media Has Vital Role in the Revolution" ("Media," 1979, p. 1).

As part of their move toward "state capitalism" in Guyana, the government of the newly declared "Cooperative Republic" moved to nationalize key assets. These assets included much of the existing media including purchase of the then foreign-owned radio stations in 1968 and 1979 (Sidel, 1984).This state domination extended to the radio (television broadcasting did not make its way to Guyana until 1991). Lent (1982, p. 376) described something of the contradictions involved in the decision to nationalize media, particularly in the case of radio:

> The government completed its control over all media in January, 1979 when it purchased Radio Demerara. The background behind the purchase illustrates the contradictions of the government regrading national development. The government, when throwing out all foreign media in 1977, asked Rediffusion, a foreign organization, to remain with its Radio Demerara.

Government monopoly control of radio broadcasting, established under socialism, continued well past the socialist experiment with recommendations still being made for dismantling the state monopoly as late as 2002. Government officials such as Information Minister Christopher Nascimento, by the early 1980s, espoused notions of "development communication" and "development support communication" and declared that the mass media were national resources whose primary responsibility was to educate and raise consciousness. Chandising (1983, p. 64) described the nationalization of the existing media as a narrowing of the democratic process under the PNC government, writing that:

> It has successfully monopolized the two means of mass communication: newspapers and radio. Both radio stations and the single national

daily newspaper are government controlled. A variety of administrative methods have been used to prevent the printing of opposition newspapers. Chief among these is the withholding of import licenses for machinery and newsprint.

Hope (1985, p. 65) similarly noted that the Guyanese government continued its hold on both publishing and broadcasting:

> Guyana's three radio stations and daily newspaper (Guyana Chronicle) are owned by the government and used in a crude manner to transmit government propaganda. News of the political opposition is almost non-existent. They are denied access to the national media and they are frequently denied permission to hold public meetings. The opposition papers (The Mirror of the PPP; Dayclean of the WPA; and the Catholic Standard of the Catholic Center) have frequently been denied permission to import or accept gifts of newsprint to facilitate their publication.

The Guyanese population was not completely isolated from contact with the outside world, with radio signals being accessible from abroad and a fair amount of trading with its neighbors in the Caribbean bringing both news and printed materials whether through legal or illicit channels. Government thus needed to maintain an internal propaganda effort to justify its actions. The government-owned and -controlled paper argued, based on the statements of one Comrade Kester Alves (described as a special political assistant to the prime minister) that:

> The media in colonial Guiana was in capitalist bondage, prevented from uncovering defects in the colonial-capitalist socio-economic system. But the media in revolutionary Socialist Guyana is not in bondage. It is free, and asked not to cover up internal social defects and not to issue misleading reports aimed at making things seem better than they are. . . . We have balanced the role of the media in colonial Guiana as a bondsman of the imperialist-capitalist system, against the free, unfettered, constructive and critical role which the media in Guyana today is expected to play. ("Free and Unfettered," 1979, p. 6)

Guyana's minister for state information told an international conference in 1980 that "a major priority of mass communication in countries like Guyana, is to realize its true potential and play a proper role in the development thrust to create a better life to the people; promote awareness of Guyanese history and cultural heritage; link its cities with the hinterland; stimulate farmers to produce more from the land; and to stir young minds to fully tap the natural resources of the country" ("Mass Communication," 1980, p. 6). This is not to say that the quest for free speech was forgotten. In the government's very insistence on media for national development can be found evidence of the tensions with private media interests during the socialist experiment. In 1983, for example, when the government sought to muzzle private publishers by denying them licenses to import newsprint, a minister in the Office of the Prime

Minister, Yvonne Harewood-Benn, insisted that "the national media must be owned, managed and operated to serve the public interest" while defending the government's actions to disallow even donations of newsprint to private media, saying that:

> For any government to allow gifts of newsprint to privately-owned media would be to open the flood gates of external intervention in the national and internal affairs and we cannot make any apology for refusing to condone this. ("National Media," 1983, p. 10)

In most other cases, including that of radio broadcasting, government control was somewhat more subtle, but often counterproductive, failing to achieve its development goals by failing to function effectively. Sidel (1984, p. 497), for example, argued that:

> There was no official censorship of broadcasting, but close listening revealed little criticism of the government, little attention to the opposition, much to socialist countries and President Burnham, and more than occasional comments that reflected unfavorably on Western developed nations. A former Guyana broadcasting official, Ron Sanders, summarized the results of such policies. "The primary effect seems to have been a loss of credibility, rendering the media virtually ineffective as either a development or information tool."

Guyana was not alone in its socialist experimentation. Among the former British colonies, Jamaica elected an avowed socialist in Michael Manley during the 1970s and touted its developing ties with Communist Cuba for several years. Guyana was also not alone in its push to nationalize its media resources.

COLONIAL AND POSTCOLONIAL RADIO
IN SOME OTHER DEVELOPING NATIONS

Mosime and Mhlanga (2016, pp. 56–57) commented on the deep institutionalization of colonial ideas through radio, writing that under colonial rule "broadcasting systems, in particular radio, functioned as an appendage of the ruling elite and were often understood within the official lenses of the colonizer" with radio serving as "a medium of communicating colonial ideas and policies to the colonized." Emerging out of such an experience, developing nations wrestling with the vestiges of colonial rule were often apt to turn inward, focusing on local mass media production as they perceived this to be a strategy to at once enable their own voices and counteract foreign influences including those of former colonial powers. Radio was an important player in the postcolonial environment, particularly where television was still expensive both on the production and reception ends. These emerging societies often caught the wave of postcolonial resistance to foreign media during early indepen-

dence and insisted on domestic content to counter the lingering influences of colonialism and the continuing domination of foreign mass media. In many cases these efforts were accompanied by efforts to include
the use of local dialects and languages, particularly where these were
suppressed or denigrated under colonial rule (St. Hilaire, 2011).

The section below examines the experiences of four additional territories with colonial emergence of radio and postcolonial transformation of
the medium, two in the Caribbean and two in the Middle East/North
Africa region.

Trinidad and Tobago

The Republic of Trinidad and Tobago gained its independence from
Great Britain in 1962 after several hundred years of colonial rule. Prior to
the British, other European powers had laid claim to the two islands
including the Spanish, French and Dutch at different times (Brereton,
2007; Williams, 1962). European colonization largely wiped out the native
Taino peoples of the islands, and imported labor in the form of African
slaves and Indian indentured laborers would eventually form the majority of the population (Samaroo, 1987). In 1889 the British combined the
two islands for administrative purposes due to their proximity to each
other. As the southernmost islands of the Caribbean archipelago and due
to their strategic position just off the coast of South America, the territory
was an important trans-shipment point for colonial goods and was also a
major producer of sugar and related products such as molasses and rum
as well as cocoa (Sudama, 1979).

The bigger of the islands of this two-island republic, Trinidad, had
been an important location for wireless telegraphy signals, partly due to
its location at the southernmost end of the Caribbean chain and just north
of the South American continent, and partly due to its importance to
colonial shipping. The island was home to the chief depot of the Royal
Mail Steam Packet Company where vessels on that line stopped on their
South American routes. It also served as "the junction for all shipping to
the West Indian Colonies," and "all goods and passengers destined for St.
Lucia, St. Vincent or Guiana" were required to first stop in Trinidad
("Wireless Telegraphy in Trinidad," 1913, p. 427). For these and other
reasons (including the colonial pursuit of petroleum resources), Trinidad
enjoyed the luxury of wireless telegraph stations before many of its
neighbors. By 1913, an aging existing station was being replaced with a
Marconi 5-kilowatt station with an expected range of 350 nautical miles
("Wireless Telegraphy in Trinidad," 1913) but which was eventually
found to frequently achieve 1,500 miles ("Trinidad Station," 1915).

Despite the vibrant wireless tradition, however, colonial authorities
were averse to pursuing audio radio broadcasting in Trinidad and Tobago for many years. Indeed, even during the 1930s, as radio burgeoned in

places like nearby British Guiana and Barbados, the colonial authorities actively suppressed amateur efforts at broadcasting in Trinidad. These efforts extended even to radio listening with the imposition of a tax on radio sets of $2.40 in the 1930s; failure to pay the tax resulted in confiscations of radios (Matthews, 1940).

Active suppression of local broadcasting limited the signal choices for radio enthusiasts in Trinidad and Tobago to direct or relayed signals they could intercept from other colonies or from further afield. The descendants of Indian indentured laborers in Trinidad, for example, welcomed Indian music from British Guiana's broadcasters as well as content from All India Radio (Mohammed, 2017). Radio listeners were also tuning in to shortwave signals from the BBC and from US stations including KWID in San Francisco, WRUL from Boston and KGEI from Belmont, California ("Radio Programme," 1940).

Starting in 1941, the United States and Britain agreed to the US establishing several bases throughout the Caribbean to support their war efforts. One of those bases was at Chaguaramas in Trinidad. In 1943 US armed forces set up a broadcasting facility on the base with the call letters WVDI, primarily as a conduit for news and entertainment for US personnel stationed in Trinidad during World War II. Given its mandate, it was not surprising that WVDI's programming consisted primarily of relays and rebroadcasts of American radio programming with some thirteen hours daily of American radio network shows, news, music and sporting events (Reed, 1945) including serials such as *The Lone Ranger*, *Roy Rogers* and *Hopalong Cassidy* ("Naps' Many Famous Faces," 2014). Despite this heavy foreign influence, local talent from Trinidad did make its way onto WVDI's air including several local broadcasters who found jobs as announcers at the station and the station's broadcasts of local calypsonian's recordings and live calypso shows (Munro-Smith, 2004; Neptune, 2007).

In 1947 the colonial government granted a commercial franchise to the Trinidad Broadcasting Company Limited operating as Radio Trinidad with call letters VP4RD, which went on the air in August of 1947 from studios in the capital, Port-of-Spain. At the helm were a former BBC presenter and producer, William MacLurg, and a former member of the ZFY management in British Guiana, Gerald V. De Freitas.

Radio Trinidad's opening broadcast on Sunday August 31 featured an address from Governor John Shaw, greetings from other stations in the region and local musicians and singers. It also aired a BBC special edition of their program *Caribbean Carnival* called *This is Trinidad* with music and other content from the colony (Boord, 1947). The station's regular programming featured some local features mixed with a heavy diet of BBC transcriptions and rebroadcasts and some US content ("Trinidad Station," 1947; "Trinidad's Commercial Station," 1947). *Short Wave News* in 1949 ("Monitor," 1949, p. 260) reported receiving Radio Trinidad's rebroadcast of BBC's *Variety Bandbox* (a variety program of comedy sketches and musical performances) as well as sponsored programming. During the 1950s and 1960s Radio Trinidad established itself as a fixture among distant listening enthusiasts in the United States (Fargus, 1953; "Hey Dad, Remember Trinidad?," 1985).

Trinidad and Tobago, even though it had no serious pretensions to socialist leanings like its regional neighbor Guyana, moved to nationalize some radio stations and its lone television operation after independence in 1963. Thus by 1968, the government announced "steps to prevent foreign companies from controlling its newspapers, radio or television stations" including wresting ownership of Radio Guardian from "Canadian-born newspaper magnate Lord Thomson" ("Trinidad Acts," 1968, p. 4).

The Trinidad and Tobago government also funded efforts such as the Schools Broadcasting Unit under the Ministry of Education, investing in radio as a form of educational support long after such efforts had been abandoned in developed nations. This particular effort dated back to 1960 and included involvement from the BBC in establishing the service.

These efforts at nationalized media would eventually crumble under the weight of the costs of these operations and their relative inability to compete with commercial ventures post-liberalization in the 1990s. Virtually all the government's media assets would eventually be divested into commercial operations with radio becoming an extremely diverse and competitive industry featuring somewhere in the region of 39 stations including ethnic, talk, sports and various music formats in operation with a fair amount of entries, exits and ownership changes over time (Telecommunications Authority of Trinidad and Tobago, 2016).

Jamaica

Jamaica gained independence from Great Britain in 1962 a few weeks before Trinidad and Tobago. Both countries had been part of efforts to form a political and economic federation of states as they emerged from colonial rule. As Lewis (1999) has documented, these efforts failed after contention about leadership and a Jamaican referendum that opted for individual independence. Jamaica's colonial past included Spanish rule (from 1494 until 1655) and British colonial occupation (until independence in 1962) and a focus on sugar as a primary export (Hauser, Delle & Armstrong, 2011). Famous today for its music and tourism industry, Jamaica was an early entrant into the world of radio.

Early radio listening in British colonial Jamaica involved the expected mix of BBC programming and intercepted US fare such as KDKA from Pittsburgh. During 1931, listeners from Jamaica wrote to Westinghouse's W1XAZ to report that they had received programming from that source ("Station Notes," 1932). Bronfman (2016, p. 60) has noted that "as more radio stations appeared in Britain as well as other parts of the Caribbean, a variety of individuals and organizations began to pressure the colonial state to build a station in Jamaica" but the local colonial authorities refused for many years.

The colonial Jamaican government first became involved with amateur station VP5PZ in 1939 at the behest of its operator John F. Grinan (described as a "well-known estate owner") and began using the station for broadcasts on November 17 of that year (Boord, 1947, p. 64). Bronfman (2016) has pointed to the fact that the government's change of heart had much to do with the role that Grinan and other amateur operators played in helping government officials resist a wave of social unrest in 1938.

The call sign of the station changed to ZQI (sometimes given as ZQI, the Government Broadcasting Station) in 1940 and broadcasts continued for about ten years until the formation of the Jamaica Broadcasting Company on July 9, 1950, when ZQI became the Radio Jamaica and Rediffusion Network, operated as a private company alongside JBC Radio (Dunn, 2014).

Radio content at the start of commercial broadcasting became heavily local even in the pre-independence era even though, in some cases, there were decidedly imperialist overtones (such as Alma Mock Yen's program *Tea Time*). ZQI officials described their programming efforts as including "the best local talent . . . properly interspersed with discourses on subjects of public interest and given by Jamaican authorities" (Boord, 1947, p. 64) and, by 1941, were claiming "a large number of reports from listeners throughout the West Indies, Central America and the eastern part of the United States and Canada" (Morrison, 1941). Bronfman (2013, p. 146) cast a somewhat less rosy light on the subject, arguing that, after years of neglect, local Jamaica radio broadcasting left a lot to be desired at the start, with only a few hours of programming daily and a "sparse menu that included classical music, educational lectures such as *The Banana* or *Typhoid Fever*, notices and news relayed from the BBC" even though owners of receivers well knew that they could listen to broadcasts from all over the world. Bronfman, however, has (notably) taken no issue with the importance and popularity of cricket broadcasts at that time.

Station management at the early ZQI operation also described the importance of cricket as a driving force for regional and international broadcasting. To facilitate more effective cricket coverage, the Jamaicans experimented with regional relays, facilitated in part by the local telecommunications operator and other regional broadcasters. One experiment in particular took game commentary and relayed it through several links to the radio system in Trinidad for rebroadcast to their local audience (Boord, 1947). Despite its low power, ZQI, in keeping with the general awareness of global reach, even programmed a small portion of its material "in Chinese from time to time" (Boord, 1944, p. 54).

The notion of radio as an educational tool persisted in Jamaica with the establishment of the Radio Education Unit at the University of the West Indies at Mona in Kingston, Jamaica. Established in 1954 and closed in 2013, the unit was meant as a tool for educational content production that would be used in Jamaica but also distributed throughout the English-speaking Caribbean as part of regional distance-education efforts (Mock Yen, 2002; The University of the West Indies, 2013). Its production efforts over the years have included a wide array of programs ranging from weekly spots on diabetes to co-productions of radio dramas on teen pregnancy and features on infant and maternal nutrition. The government's Jamaica Information Services (JIS) also developed a system of schools' broadcasts to serve educational needs.

Radio in the post-independence Jamaica was at once a source of national pride and a bone of contention. On the one hand, radio was marked by vibrant public discourse and debate, on the other, it was plagued by the notion of residual colonialism. For many years, critics of the Jamaica Broadcasting Corporation, for example, railed against its BBC-fashioned public broadcasting model and its broadcasts of BBC programming including such radio dramas as "A Judgement for Julia" and "Here Comes Charlie" and musical titles like "Festival of Waltzes" and "Concert Hall" in the 1960s (Gordon, 2008, pp. 31–32). Later, when the JBC's FM frequencies insisted on programming only classical music, that decision promoted further suspicions of elitism if not lingering colonial values. It would be several more years and well into the late 1980s until the government began to liberalize the mass media and allow further private licensing of radio stations that included newcomers such as KLAS radio and, later, in 1990, Jamaica's first all-reggae station, Irie-FM (Mohammed, 2017).

Algeria

Algeria's fight for independence resulted from the French colonization of the territory starting with the French invasion of the North African territory that was a regency of the Ottoman Empire in 1830 (Sessions, 2011). In 1962, Algeria gained its independence from France after several years of violent clashes between local rebels and French authorities. Its early independence, however, was marked by a repressive socialist regime and several subsequent political upheavals (including civil war) have marked its history as well. Its linguistic traditions include Arabic from early Arab settlers and a continuing French influence alongside the native Berber Kabyle language. Algeria's geographical location as part of Africa very close to Europe has made its radio history particularly interesting, as has its use of radio in its struggles for national identity. For Scales (2013, p. 305), "the emergence of radio broadcasting transformed the dynamics of colonial politics in French Algeria by making broadcast sound into a new site of struggle between Algerians and the French colonial state."

Owing to their country's proximity to Europe, early radio listeners in Algeria tuned in to a wide range of European broadcasts, many of which by the 1930s were even producing content in Arabic or featuring music from the Middle East and North Africa region. Algerian listeners, for example, in 1928 congratulated Radio Toulouse from France on their program of Moroccan content ("Dominion and Foreign Broadcasting," 1928). Earlier, in 1924, an Algerian radio enthusiast sent a letter to the BBC indicating that listeners were enjoying their Chelmsford broadcasts in Algeria ("Algeria Hears Chelmsford," 1924). Yet, particularly as not all the content being received was of French origin, this radio listening prompted concern among colonial authorities as Scales (2010, p. 385) has outlined:

> During the 1930s, transnational radio broadcasting fueled fears within Algeria's colonial administration, France's security services, and in metropolitan political circles about the conjoined threats of foreign subversion and domestic political instability in the colony. The rising Fascist powers of Germany and Italy challenged France's geopolitical status on the Continent and in its colonies with their newly streamlined radio propaganda machines.

These tensions prompted interest in local broadcasting particularly as a counter, in the first instance, against extra-colonial influences. Aïtel (2013) placed the start of Radio-Alger in 1925 with broadcasts in French. As early as 1931, radio broadcasts from Algiers included a series of talks in English on the subject of "Algeria" ("Algiers Correspondent," 1931). In 1942, distant listeners were reporting that Radio-Alger, which was affected by wartime developments, had gone from regular to sporadic broadcasting but could still be received with greater strength than their neighbors in Morocco (Morrison, 1942). Following the end of World War II, Radio-Alger formed a division known as ELAK (*Émissions en langue arabe et kabyle*) that gave rise to Chaîne 2 in 1948, a channel devoted to broadcasting in the local Kabyle dialect associated with the Berber ethnic group.

Postcolonial Algeria inherited the infrastructure for mass media (including radio) left behind after the war of liberation ended with the expulsion of colonial French rulers in 1962. Leaders and media practitioners in the former colony viewed the facilities with some ambivalence. On the one hand, these were former tools of colonial domination, which for their continued programming would require imports of resources and content. For Chevaldonné (1988, p. 270), "the logic of the system which had been left in place tended, without draconian counter-measures, to work in a neocolonial fashion." Conversely, they also held the potential for creating an industry and a voice nationally, and to bolster communications and program sharing with regional countries rather than foster dependence on colonial powers (Chevaldonné, 1988).

After independence, the new government nationalized the existing media facilities under the title of *Radiodiffusion télévision algérienne* (RTA) on October 28, 1962. During the years following independence, the National Liberation Front (FLN) government heavily controlled the nationalized resources of the former colonial media systems, of which radio was the most important (reaching some three million people domestically) despite a supposed constitutional guarantee of freedom of the press (Walpole et al., 1965).

Radio broadcasting continued with a system of three radio networks developed under French rule with each focusing on a different language—French, Arabic, and the local dialect, Kabyle—but with some level of foreign content including "classical, folk and jazz musical programs from the United States, France, Italy and Communist China; discussions from Warsaw and Moscow radio," French dramas and discussions of African history (Walpole et al., 1965, p. 336). This three-pronged system, however, masked a system of what has been termed *"Arabisation"* (Arabization) in which the state used radio (along with television and other media) to promote the use of classical Arabic as a *lingua franca* to the exclusion of French while also marginalizing the native Kabyle (Aïtel, 2013; Grandguillaume, 1998). This strategy was eventually rejected with more recent efforts to recognize the local dialect as an official national language.

The interplay of radio in the processes of postcolonial independence and the concurrent process of national and regional identity in places such as Algeria are evident in the work of Frantz Fanon (1970), who pointed out that Algeria's emergence out of colonialism was part of a process that included not only the spread of radio in Algeria but also the spread of broadcasting in its neighbors including Syria, Egypt and Lebanon.

In some cases, the impetus to invest in broadcasting had as much to do with economic factors as with concerns about colonial influences and national identity. This was at least partly the case in the Arab Gulf countries where, according to Boyd (1999a, p. 5), economic opportunities, external propaganda threats and domestic communication needs led to investments in broadcasting:

> Those countries that were initially slow to develop radio services—Kuwait, Saudi Arabia, Qatar, and the United Arab Emirates—were also most vulnerable to radio propaganda from countries such as Syria, Iraq, and Egypt, which had political interests in deposing the Gulf ruling families. In the 1960s and, of course, the 1970s, these countries rapidly developed radio broadcasting facilities because they recognized the need for reliable communication with the indigenous population as well as with the Arab expatriates who came to work among

them. The oil-rich countries could afford modern high-powered transmitters and impressive radio production studios.

Elsewhere in the Middle East and North Africa region, radio had taken root earlier under colonial and then national government control.

Egypt

Egypt's documented history spans more than 5,000 years. Its several main historical influences have included the ancient Egyptian culture, the influence of Islam under the caliphates of immediate successors to Muhammad, as well as Ottoman rule and French and British occupations. The modern state of Egypt emerged through several stages of British withdrawal from colonial rule including being assigned the status of protectorate in 1914 and Britain's declaration of Egyptian independence in 1922 (Marsot, 1985). However, continued British presence and involvement would not end until a revolution in 1952 when Egyptian military officers deposed the monarchy and expelled the British (Tignor, 2010). Egypt's position in North Africa and its ancient status as a center of trade and commerce also placed it among the early crossroads of radio wave traffic, as one observer put it, between the East and the West.

Egypt had been part of wireless efforts prior to the development of audio radio systems. In 1922 the British government opened the Abu Zabal wireless station outside Cairo. This station was meant to be part of the British proposed imperial wireless chain but changed hands to Marconi after plans for that system changed ("Wireless in Egypt," 1926). Sporadic efforts to start radio were underway in Egypt by the 1930s but the economics, even under British supervision (and the license-fee model that they had backed at home and elsewhere), were problematic and attitudes to the new medium were mixed. An early report in a 1933 journal ("Current Topics," 1933, p. 195) recorded the skepticism expressed in the local newspaper *Al Sh'aab* about broadcast radio, suggesting its imminent demise despite the existence of some 50,000 receiving sets:

> The radio set is an instrument for the connection of thought between the East and the West before it is an instrument for amusement and joy. It enables one to hear lectures on various subjects, and to follow the development of thought in all parts of the world. But do all Egyptians who have radio sets in their houses know how to derive the desired benefit therefrom? . . . And do they care to listen to a lecture given in French or English or even in Arabic? It is almost certain that many of the Egyptians who have radio sets in their houses never use them except only to reproduce the voices of Amm Kalsoum,* Mohammed Abdel Wahab (well-known Egyptian singers) or the monologue of the stupid pupil, etc. In other words, they use the radio set as a phonograph. (*أم كلثوم, more commonly rendered as *Umm Kulthum*)

Settel (1938, p. 37), writing in the international press, had a somewhat more enthusiastic (if condescending) notion of the impact of radio on Egyptian society, writing that the new medium had, by the late 1930s, displaced more traditional forms of entertainment such as coffee house sideshows:

> Today, from the smallest and most obscure coffee shop one can hear the ear-drum-busting wail of native songbirds squawking from the Egyptian Broadcasting Co. studios in Cairo's business center. A handful of stars plus payroll people who read out the Koran (Moslems' Bible) monopolize the programs. . . . Ye Olde Moslem Sheikh, decked out in all his glad rags, squats over the nargilla and listens to Cairo's crooners. . . . Coffee-house proprietor, spared of having to foot the bill for a lot of 12th-rate actors and feeding them to wit, thinks the radio is the greatest blessing since the Prophet and he gives music plenty to his customers.

A somewhat derisive report in the mid-1930s described the Egyptian radio efforts to make an international impression and hinted at their lack of funding:

> Egyptian government is resorting to radio for tourist baiting purposes and is having a series of spiels waxed. These will be sent around and put on wherever Egypt can make a deal with radio stations in Europe, South America and elsewhere. Government doesn't like the idea of paying for the outlets, so is offerings the plates around on a reciprocal basis. Offers to devote time on Egyptian radio to tell about the beauties of Poland or what have you in return for time abroad. ("Egypt Heard From," 1935, p. 41)

The Anglo-Egyptian Treaty in 1936 granted Egypt full independence. Its new ruler, Prince Farouk, was known to take to the radio airwaves to deliver what were characterized as "occasional popular radio broadcasts addressed to his devoted people" ("Egypt Comes of Age," 1938, p. 48). This direct government role would be consolidated in varying degrees over the history of Egyptian radio (Boyd, 1999a).

Egyptian radio interests faced problems of few receiving sets during the 1930s, which limited the usefulness of receiver license fees in helping to develop or maintain broadcasting. In 1937, the Egyptian government, having committed by that time to radio broadcasting under official control, resorted to a scheme of providing as many as 2,000 radio receivers free of cost in selected villages throughout the country intended to bring radio including educational broadcasts to the rural (and often illiterate) populations ("Radio in Egypt," 1937). This initiative, called ambitious in some international press reports, also raised fears that the Italian Fascists might be able to cut in to the broadcasts and so influence the Egyptian listeners against their own government ("International Radio: 400," 1938).

Eldin (1941, pp. 34–46) noted that among the rural listeners were Egyptians who lived in the most primitive conditions and often depended on the distributed radios for their news, with the anticipated problematic results for Egyptian and British authorities:

> Their only source of news, since they cannot read, is the radio. At sunset many of them gather at the village cafe, squat under a fuming oil lamp, and play game after game of backgammon, stolidly smoking their *narghilies* or water-pipe. The appearance of the newscast is greeted by respectful silence and the suspension of all activity. Unfortunately, the news thus heard often comes from the Italian station of Bari, where a renegade Iraquian [*sic*] speaker broadcasts vitriolic anti-British propaganda supplemented by daily announcements of crushing British defeats.

This situation prompted the BBC to start broadcasts to the region in 1938 with news and entertainment in Arabic to appeal to the same audiences the Italians were targeting.

Still, even as Egyptian radio broadcasting struggled for funds, and battled foreign influences, Egypt envisioned its eventual development into a regional radio broadcasting force, and plans were already in place (though funding continued to be a problem) for regional shortwave reach to the Near East and Sudan by the early 1940s ("Radio: Egypt Station," 1940). Meanwhile, its neighbor to the south, Sudan, was planning for its own radio broadcasting facility around the same time, with the input of Sudanese, British and Egyptian staff.

For all their efforts to support radio, later national governments would also exercise strict control over broadcasts. An Associated Press report in 1957, for example, indicated that all radio broadcasts in Egypt were subject to prior approval from an official censor ("Egypt Censors Radio," 1957). While information was being controlled at home, Egypt felt the need to expand its broadcasts abroad with the establishment of their external service Voice of the Arabs on July 4, 1953, out of Cairo.

Criticized as a propaganda tool of the Nasser regime, the Voice of the Arabs service enjoyed wide reach across the region, reinforcing regional ideas and taunting what they saw as imperialist forces such as the British (Hale, 1975). British news sources characterized the station's programming as a "stream of abuse" ("Egyptian Radio Propaganda," 1955, p. 7) against the West, broadcasting for several hours a day in Arabic and Swahili with some programming in German and Hebrew aimed at Israel.

Some of Egypt's neighbors also blamed the Voice of the Arabs station for engaging in open propaganda efforts. In 1955, amidst tensions between Iraq and Egypt, the London newspaper *The Times* reported that the station had repeatedly attacked the Iraqi prime minister and his government ("Radio Weapon," 1955). To complicate matters, two other stations could also be heard hurling propaganda into the airwaves around these political tensions, including the Voice of Free Iraq (said to be broadcasting from Egypt) and the Voice of Free Egypt (an anti-Nasser operation whose location could not be determined—but in which some suspected the hand of the British) ("Radio Weapon," 1955).

REGIONAL NEWS AND RADIO IN DEVELOPING NATIONS

While local media production was often the key element of their strategies for information self-reliance, many (particularly small) states also explored the possibilities of regional cooperation and exchanges among fellow developing nations and contiguous countries. Chief among their reasons for so doing was the perceived imbalance in international news sources. According to Martin (1988, p. 71):

> When Third World spokesmen and sympathizers demanded a change in the manner and content of reporting about their affairs, a number of end products of these demands were envisaged. At the top of the list of priorities was the establishment of their own national and regional news agencies. These agencies would satisfy broadly two desires—one to end the dominance of Western news agencies in reporting their affairs, and the other to add to the range of institutions which have frequently characterized the entry into nationhood.

Efforts at program exchanges and the development of regional news agencies were sporadic and enjoyed varying degrees of success. These included such efforts as the Caribbean News Agency (CANA), formed in 1976, and the *Acción de Sistemas Informativos Nacionales* (ASIN) in Latin America and the Pan African News Agency (PANA), both established in 1979. These were, in their initial stages of development, largely press-type wire services supplemented by periodic radio broadcasts. The Caribbean News Agency, however, became a regional player in radio with the launch of CANA Radio in 1984. CANA Radio boasted as many as 17 subscribers by 1988 and featured varied programming including daily news reporting packages with "The Caribbean Today" (usually run as a segment within or after major local newscasts on domestic stations) but also featuring more ambitious content such as regional interconnections among subscribing stations. Brown (1995, p. 307) described CANA Radio as being a primary driver of the Caribbean News Agency's popularity in the region:

> The prominence of the Agency within CARICOM is due in large measure to its radio service. CANA Radio, which started operations in 1984, originates a daily 15-minute news program—The Caribbean Today— which is carried in the majority of CARICOM countries. . . . The program is carried live by some stations but is recorded and re-broadcast by others. CANA Radio also produces and distributes a variety of other news and public affairs programs which are carried by subscribers

both inside and outside the region. A once-monthly live discussion program on a topical matter—Crossfire—is also a popular CANA Radio production.

Produced in Barbados, CANA's radio content provided a regional view with information on developments of interest to the Anglophone Caribbean and news of major events in individual territories. These territories had chosen individual independence from Great Britain over formation of a federation of states. In the post-federation era, exchanges of news were rare enough as the various former colonies focused on their individual pathways to development without the prospect of a political union. Regional contact and cooperation were reduced to the realms of trade agreements such as the Caribbean Free Trade Agreement (CARIFTA), established in 1965, and its successor, the Caribbean Community (CARICOM), first established in 1973. While these various trade agreements served to unify broad policy objectives with regard to regional and international trade, they did very little to integrate the various populations of the region. A colonial holdover, the game of cricket perhaps did far more to maintain a regional consciousness than any trade agreement as the islands of the Caribbean and Guyana in South America continued to play against each other and to compete as a unified team—"The West Indies"—in international cricket.

With its daily contributions of regional news, the CANA Radio service served to highlight regional issues, inform about major developments and—at the same time—maintain the subtle variations of Caribbean dialects and accents among its listeners throughout the islands and in Guyana. In this regard Myles (2000, p. 107) noted that voice, dialects and accents are important sociological concepts that impact radio and define its roles in society since "speech, when it becomes identified as the articulation of difference in dialect or accent, is a form of voice that claims recognition through the agency of cultural forms like music and talk radio." CANA Radio helped to maintain a broad scope of reference for Caribbean "voice" on radio in the region that evolved through many stages in later years into a variety of program and content exchanges on radio and television. These systems, perhaps without often recognizing it, demonstrated something of a unifying legacy of British English inherited from colonial rule and spoken in its many local variations in the Anglophone Caribbean region.

Cuthbert (1990, p. 410) agreed with the suggestion that CANA Radio was influential, adding that its ability to link stations from separate territories and address issues of regional interest was a particularly strong draw:

> One of the CANA Radio programs, "Caribbean Crossfire," provides live linkups once a month for all interested stations, and has dealt with topics such as cricket, AIDS, CARICOM, calypso, Jimmy Swaggart,

regionalism, drugs, industry, sugar and tourism. As many as 14 stations have been linked at one time for these programs. CANA also had
a special linkup program called "Haiti Watch," when Haiti was much
in the news.

As late as 2002, efforts at regional radio collaboration were continuing
as members of the Organization of Eastern Caribbean States announced a
nine-station radio news exchange on March 4 of that year:

> Radio Anguilla, ABS in Antigua, ZBVI in the British Virgin Islands,
> DBS in Dominica, GBN in Grenada, ZJB in Montserrat, ZIZ in St. Kitts/
> Nevis, RSL in St. Lucia and NBC in St. Vincent & the Grenadines each
> week-day contribute stories to a 15 minute news program called
> "News- link." The package of regional stories is then broadcast during
> their late afternoon programming. The news exchange is being coordi-
> nated by the OECS Secretariat with support from Radio St. Lucia.
> ("OECS Radio," 2002, p. 3)

So that while the Guyana case and several others provide examples of
the turn to nationalist media, they also amply demonstrate the transformation of radio from international to national and regional in many developing nations. As we shall consider later, however, this dalliance with
introspective and nationally focused media to the exclusion of outside
influences would, in varying degrees and paces, but with inexorable cer-
tainty, fade in the face of globalized information technologies in later
decades.

With changes in technology, also, came private and commercial efforts that sought to develop local and regional radio as profitable business enterprises. Early efforts at commercial regional Caribbean stations
included Gem FM; launched during the 1990s (originally out of Montser-
rat), it focused on providing programming to several of the smaller islands with a regional scope. Gem FM moved to Antigua and ceased to
operate under that name by 2014. The possibilities of reaching larger
audiences continued to be attractive, and media companies have made
several efforts since then to regionalize their radio broadcasts, particularly as Internet connections made audio links cheaper, more accessible and
higher quality than satellite or telephone links.

The commercial "Caribbean Super Station," for example, launched in
April 2011 and broadcasts to nine of the Anglophone Caribbean islands,
the majority of which are separate independent countries. Despite being
a commercial, for-profit business venture, the station frequently express-
es a commitment to particular political ideals such as regional integration
and Caribbean community (One Caribbean Media, 2018). It pursues these
ideals, however, through shared music, entertainment and sport rather
than overt political activism.

SIX

Radio, Changing Technologies and Changing Roles

While radio in the United States underwent transformations related to deregulation and corporate buy-ups of local stations, the age of nationalized radio in many developing nations was beginning to unravel. During the late 1980s and early 1990s, many developing nations were beginning to re-think their strategies of government ownership and control of mass media. Not only were the costs of running nationalized media growing increasingly onerous, but private interests were also clamoring for the freedom to broadcast and, where given the chance, were doing so more efficiently than government broadcasters. In several places, governments relaxed long-standing bans on private broadcasting and granted licenses to private companies.

Storr (2011, p. 554), writing about the English-speaking Caribbean, argued that, while "public broadcasting evolved in the English speaking Caribbean as state broadcasting," forces such as deregulation, liberalization and privatization placed pressures on governments to open up broadcasting, particularly in the late 1980s and 1990s. Under the financial strains of running television and radio operations, the Jamaican government, like their counterparts elsewhere, began to issue broadcasting licenses to general purpose radio stations like KLAS FM (which would eventually turn into a sports station) and specialist stations like Irie FM— Jamaica's first all-reggae radio station.

As private stations emerged in many developing nations, they did so not only with less expensive studio and transmitting equipment as stations became digital and automated, but also with various niche audiences and special interests in mind. Small independent stations found that they could focus on segments of a population (rather than a national audience) even in relatively small countries. Stations were also able to

111

function outside of national mandates that would otherwise dictate their content, so that experimental stations and innovations became possible.

In other developing nations, governments were under less pressure for privatization and deregulation and, rather, faced challenges that re-quired novel uses of state-owned media resources. This new wave of radio activity, in the transition from state-owned and -controlled media to private media, often featured combinations of private and publicly owned broadcasting or even re-deployment of old broadcasting technol-ogies to serve emerging needs. One important example of how this new wave of radio (even where governments still funded radio operations) combined with new technologies has stood out—the so-called "Kothmale Project."

KOTHMALE, RURAL RADIO MEETS THE GLOBAL DIGITAL

The Kothmale example arose out of a Sri Lankan government project in community radio called the Mahaveli Community Radio system in which low-powered transmitters served small communities with volunteer staff running the stations. World Bank documents described the Kothmale Community Radio Internet Project as "an attempt to extend the benefits of information and communication technology (ICT) to some of the re-mote areas of Sri Lanka through the innovative convergence of two me-dia—the radio and the Internet" (Bhatnagar, Dewan, Torres & Kanungo, 2003, p. 1). Seneviratne (2011, pp. 130–131) described how one of the stations in the system, the Kothmale Community Radio (KCR) station, pushed the envelope of broadcasting practice by linking digital net-worked technologies with more accessible radio:

> In 1998, UNESCO provided US\$ 50,000 to start implementing a project, which came to be known as the Kothmale Community Radio Interorg Project (KCRIP). . . . The project, designed to address the problem of rural access to computers and connectivity, was set up as a mini-ISP with leased line connection to the Internet and seven computers—three for public access at KCR, one in the studio, one server, and two for remote access from public libraries in neighboring towns.

As Seneviratne (2011) has noted, this project gained international attention for its novel combination of radio broadcasting and the then relatively new technologies of the Internet and the World Wide Web, particularly in the context of a rural area in a poor developing nation. For Seneviratne (2011), the project was also notable because it demonstrated how the accessible technology of radio, broadcast in a local language, could make relatively inaccessible technologies (such as computers and the World Wide Web) available by breaking both technical and linguistic barriers. Broadcasters used the Internet to research their radio produc-tions and to respond to listener queries (in local languages) about diverse topics in a process that came to be known as "radio browsing the Internet" (Seneviratne, 2011, p. 131).

The Kothmale project reflected the diverse and changing roles of radio (particularly in its relationships with emerging digital, networked technologies) that could be observed, particularly outside of the United States and other developed nations. Radio has served alongside globalized digital networked technologies, at times, to provide communities with proxy access to the digital global, often simultaneously overcoming linguistic barriers to access.

This interplay of radio with the digital global has taken diverse and novel forms in the context of audiences and communities outside of the digital mainstream. As part of the Kothmale project, for example, digital content production collectives and social groups such as listeners' clubs and Internet clubs have been formed, comprising school children, university graduates, vocational trainees and others (Bhatnagar, Dewan, Torres & Kanungo, 2003).

RADIO AND THE (DIGITAL) SOCIAL

Communal radio listening has been a feature of radio audiences since the early diffusion of sound broadcasting and one that has persisted in various forms. Goodman (2016) has noted that listening groups were popular in the United States and the United Kingdom in the early days of radio, and persisted, with the involvement of the Australian Broadcasting Commission (ABC), into the 1950s in Australia and with Canadian Broadcasting Corporation (CBC) efforts in Canada into the 1960s.

At times, communal listening has been a solution to the problems of the cost of radio receivers and the technical issues of providing power to those receivers. *Radio Craft Magazine* ("India's New Network," 1938), for example, described how the research staff of All India Radio developed special receivers intended for communal listening in villages. In other situations where early radio technology was too expensive for individual ownership, group listening was similarly common, as is evident in the noise restrictions imposed on radio listeners in British Guiana who were in the habit of listening en masse to the loudspeakers of whomever had the means to purchase and operate a receiver. Outside of the strict influence of cost considerations, group listening has also been the result of social circumstances. In cases such as Franklin D. Roosevelt's fireside chats, communal listening reflected solidarity of concern and an attempt at appealing to a national community facing shared challenges.

Yet, group listening as a function of cost would not disappear, even as transistors drastically reduced the size and costs of radio receivers. Group listening has continued to be an important strategy for reaching audiences in certain cases. In the previously mentioned 1990s Tanzania-based HIV/AIDS education project *"Twende Na Wakati"* (Let's Go with the Times) project, for example, radio was the primary medium for dissemination of the audio drama content that aimed to educate audiences about the disease, addressing misconceptions and encouraging prevention behaviors. When workers from the Tanzanian Population and Family Life Program (POFLEP) went into the rural villages that would be their target audiences, it was quickly evident that a lack of radio ownership was a barrier to the program's reach. The most cost-effective solution was to provide sets and batteries to rural health centers at which staff would tune in the scheduled broadcasts for health center attendees and others from the village to listen (Svenkerud, Rao & Rogers, 1998). Such was the audience involvement in that project that listeners were known to write letters to Mkwaju, the (fictional) promiscuous truck driver character, encouraging him to follow the advice of the program and avoid unprotected sex. These letters were sent to the characters in care of the local radio stations. Additionally, community discussions of the program and its content formed a major part of the evaluation of the project's effectiveness (Mohammed, 2001; Singhal & Rogers, 1999; Svenkerud, Rao & Rogers, 1998).

In the case of the *Tinka Tinka Sukh* (Happiness Lies in Small Things) radio soap opera broadcast on All India Radio (AIR) from 1996–1997, the program focused on gender-equality issues with a particular emphasis on the illegal practice of dowry in which grooms would demand money from the families of brides as a condition of marriage (Papa et al., 2000). Evidence of the community of listeners was found when AIR "received a colorful 30 by 24 inch poster-letter from Lutsaan village in India's Uttar Pradesh State, signed by 184 community members, pledging not to give or accept dowry" (Papa, Singhal & Papa, 2005, p. 173).

More recently, Damome (2011) described the formation of radio station listeners' clubs in several African nations, particularly as Internet access has spread, providing opportunities for listening beyond the airwave reach of the stations and greatly enhancing the possibilities of listener-station and listener-listener communication. Such clubs provide virtual and real social meeting opportunities for audience members (many of whom listen to the stations primarily online) and a source of audience feedback for station programmers. According to Damome (2011, p. 243), club members "organize demonstrations in support of the radio station, make contributions and participate in various aspects of the life of the radio station," providing feedback to station management and programming advice:

> At the *Radio de l'Alliance Chrétienne* in Burkina Faso the meetings of the club are called *"graine"* (seed). The club meets every month in the presence of the director of the station for a drink and talks about the radio station. The community of listeners of *Radio Maria* in Togo and the

Radio de l'Immaculée Conception in Bénin also meet regularly. For them, it is truly a "family" surrounding the radio stations.

Digital networked social media systems have extended the possibilities of community and interactivity for radio—elements that have demonstrated their importance to various communities of interest. The Indian radio phenomenon in Trinidad, for example, has used Facebook, Twitter, text messaging and other social media for community formation and maintenance, including providing virtual meeting places for listeners in diverse locations, opportunities for interaction with broadcast personalities, and discussion spaces for shared topics of interest (Mohammed & Thombre, 2017). Similarly, Lwanda (2014) described the evolution of Malawi radio talk programs that feature Internet chats and have emerged out of a tradition of farmer's radio clubs going back to the 1960s. These combinations of radio (whether traditional or online) with digital networked technologies result in communities of listeners defined by common interests rather than by their geographic distance from a station's transmitters as in the past. Further, these listeners/community members have also, through the interplay of radio and digital networks, transformed from passive listeners consuming a one-way stream of communication into an active community providing feedback to the stations and generating discussion among the community membership.

Rural identities, cultural identity and community are also central to the digital and social roles of radio. In a broad survey of community and participatory radio stations around the world, Gumucio Dagron (2001, p. 14) called radio "the most appealing tool for participatory communication and development," and "the ideal medium for change," noting that:

> In the mid-40s, about three decades before diversity in electronic media would spread, small and often isolated communities of campesinos (poor farmers) or miners in Latin America had already started operating their own stations, not only to challenge the monopoly of state media, but also to have, for the first time, a voice of their own.

Gumucio Dagron (2001) chronicled numerous examples of radio being used around the world for community building and identification including the well-documented network of miners' stations in Bolivia that started as early as the 1950s (Huesca, 1995; O'Connor, 1990), Kothmale as described above and Nepal's Radio Sagarmatha.

More recently, *Radio Zapotitlán* in Mexico started in 2000 as an emergency measure in response to a volcanic eruption (Hayes, 2018) and primarily focuses on its local community and surrounding villages in a small part of rural Mexico under a community broadcasting license from the Mexican government. The campesino identity of a rural peasant farmer is important in these communities, as is the fact that many of the people in these areas tend to be migrant workers who may move to the United States or to other regions of Mexico in search of labor work. For this reason, *Radio Zapotitlán* serves a social function of maintaining identity among locals while digital streams share this local identity with dislocated or diasporic members of the group. Similarly, Zambrano (2018) found that even commercial stations in places such as Colombia have evolved digital presences that encourage not just wider audience reach but also community building and interactivity.

The digital social can provide more than just opportunities for listeners and stations to create community engagement and involvement. In some cases, creative combinations and application of digital and mobile networks provide the very means of dissemination of audio content. Moyo (2012), for example, described the digital and social antics of an operation known as Radio Dialogue in Zimbabwe that emerged in the city of Bulawayo in 2001. The station described itself as a nonprofit "community radio station" aiming to "debate and discuss current political, social, cultural and economic issues affecting the community of Bulawayo" (trickleout.net, 2015). Zimbabwean authorities denied Radio Dialogue a broadcast license and even pursued legislative avenues to suppress its efforts to broadcast content. These efforts were partly to support the state-run broadcast monopoly of the Zimbabwe Broadcasting Corporation (ZBC) that assumed its Zimbabwean name in 1980 but began life as the Rhodesian Broadcasting Corporation in 1963 (Commonwealth Broadcasting Association, 1988).

The station responded with a set of strategies centered on community action and technical means of bypassing the government's restrictions. Though the government excluded Radio Dialogue from the public airwaves, the station distributed its content, including news and current affairs programs and information, privately among listeners who formed Radio Dialogue ward committees. At first, the station distributed content on cassettes and audio CDs through underground networks to avoid government interception. Listeners played the news, discussions and music privately or they might hear the programming as bus and taxi drivers played them in vehicles or as owners played it in pubs and beauty salons (Moyo, 2012). By 2009, the station started a shortwave broadcasting service as Zimbabwe Community Radio (ZCR) broadcasting into Zimbabwe first from the United Arab Emirates, and, later that year, from South Africa (Chiumbu, 2009). Later, the station employed the Internet, social media and even mobile phones to distribute programming and information, including a system known as the "interactive voice response" (IVR) system that they call the Freedom Fone (Moyo, 2012, p. 485) that supports multiple communication modalities including text, voice, email and others.

Meanwhile, the Zimbabwe government actively fought against Radio Dialogue's efforts to reach their audiences, accusing the station of promoting regime change and raiding its offices in 2013 ("Radio Dialogue Falls," 2017; "Radio Dialogue Manager," 2013), seizing 180 solar-powered shortwave receivers and arresting several staff members (charges against the editor, Zenzele Ndebele, were eventually dropped). Even when the government decided to issue licenses for radio stations some years later, they continued to deny Radio Dialogue a license ostensibly because such licenses were to be issued to commercial stations only, though press reports speculated that many of the commercial license awardees had links to the government ("Radio Dialogue Falls," 2017). By 2017 Radio Dialogue was facing financial difficulties and competition from other stations and feared that it would collapse without financial aid. Eventually, a commercial operation called SkyzMetroFM took over the Radio Dialogue facilities in Bulawayo (Zhangazha, 2018). In 2018 the Zimbabwe government announced that it would allow community radio stations to obtain licenses (Ndlovu, 2018).

Zimbabwe's radio environment also exemplifies the digital social in an international context. While Radio Dialogue operated within Zimbabwe, creating a community of listeners in Bulawayo and beyond, it also adopted a strategy of broadcasting from outside the county back into it to bypass their own government. This was a strategy that Zimbabwean radio broadcasters on SW Radio Africa and Voice of America's Studio 7 have also used to reach audiences in Zimbabwe from locations in the United States and the United Kingdom. In addition to the use of traditional shortwave broadcasts, these services have also incorporated Internet streaming and mobile phone reception into their audio dissemination strategies (Mabweazara, 2013). While many examples of the digital social in global radio focus on the global reach of communities from their locales to their diasporas, several Zimbabwean radio operations have demonstrated that the flow of information can also be reversed, enabling content from external sources to reach local communities.

While community radio stations are known to use the flexibility and power of digital technologies to promote and maintain local identities even across dispersed geographic spaces, the digital social has itself emerged in ways that may serve to redefine notions of community and identity. Turkey's *Nor Raydo* presents such an example. Algül (2013, p. 89) described this Internet radio station as broadcasting in no less than nine languages with its mission statement describing it as the "voice of multiculturalism." Under the guiding principles that none of its content should defend nationalisms or discrimination, the station accepts programming created by diverse members of its community, including minority groups that are underrepresented in the discourse of mainstream media. Thompson, Gómez & Toro (2005) similarly point to the role of Costa Rica–based station *Radio Internacional Feminista*/Feminist International Radio Endeavour (FIRE) in defining their community as a globally dispersed group of people who share a focus on female voices and issues. Both *Nor Raydo* and FIRE use digital technologies and audio media to redefine notions of communities in terms of shared interests and aspirations rather than geography and shared heritage.

AUDIO, RADIO AND DIGITAL TECHNOLOGY

While Kothmale, Radio Dialogue and more recent examples demonstrate the interplay of radio and networked digital technologies, several technical developments had previously laid the foundation for the emergence of radio as a global digital phenomenon. Hendy (2000) pointed to the processes of digitization of production and distribution, convergence with other technologies and the creation of interactivity as key categories of recent technological developments. These processes are all intertwined and made far more complex today with radio's social media integrations. The historical roots, however, often had less to do with creating a global interactive radio platform and more to do with solving specific technical problems.

Scientists and researchers during the 1960s and 1970s began to harness the emerging availability of digital computing power to solve problems in their particular fields. Among these scientists were pioneers of digital audio such as Thomas G. Stockham, who was behind a 1976 RCA release of remastered early twentieth-century recordings of opera singer Enrico Caruso. According to McLellan (2004, p. B16), "Stockham and his colleagues digitally eliminated surface noise and compensated for flaws such as the tinny sound and echoes caused by the primitive recording horns used at the time" resulting in "stunningly clear and clean restored recordings of the great Italian tenor."

While the achievement of restoring deteriorated audio recordings was important in itself, Stockham was also contributing to the evolution of radio by developing techniques to convert traditional analog sound signals (recorded on media such as vinyl discs, metal studio masters or magnetic audio tape) into computer-accessible data that could then be processed.

To put this development in context, the existing technologies around audio recording and production were initially completely analog. The audio industry undertook even relatively advanced productions such as multi-track studio recordings using wide magnetic tape designed to record and read across several magnetic audio heads at the same time. Musicians and producers achieved echoes and similar effects from physical boxes containing a speaker at one end, a microphone at the other and a set of vibrating springs in between. Musical instrument digital interface (MIDI) technology that allowed instruments and music to be computerized would not become a reality until the 1980s.

Into this environment, the notion of digitization was extremely specialist and limited to special projects such as Stockham's restoration work. Music and other audio would not be routinely digitized or manipulated on a computer. As with every other technology, several efforts and numerous factors shaped the eventual emergence of digital audio.

The ability to sample real-world audio and convert those impulses to digital data was limited in its usefulness by the size of the data and the processing power required for its storage, use and manipulation. Such limitations meant that it would be several years until the system of pulse code modulation using sampled signals found its way into industry and then to the home. Radio stations were adopting a new optical system using pulse code modulation known as the compact disc (CD) by the early 1980s. At the launch of the CD on March 1, 1983, the Dutch company responsible for the experimental broadcasts from Eindhoven back in the 1920s, *Philips' Gloeilampen- fabrieken*, once again demonstrated its penchant for innovation in collaboration with the Sony Corporation of Japan (Sykes, 1984, p. 367).

A report in the periodical *Broadcast* (Shenfield, 1983) revealed that stations in the United Kingdom during the early 1980s were excited about the much better sound quality of the CD but concerned about the cost of the units and availability of discs, with one radio executive predicting that while CDs were useful they would never be a replacement for the large libraries of vinyl discs. The BBC brought the CD to its international operations in 1984 when it released a concert recording on CD for distribution to its international affiliate stations in August of that year ("Distribution," 1984). Similarly, in the United States, sales of CD players and discs quickly eclipsed the analog records and cassette tape technologies, bringing digital audio to the radio industry and the home. Lasar (2016, p. xii) listed the compact disc among a range of "pre-Internet technologies that allowed for the individualization of listening," including the boom box and the Walkman, but described the CD as the "consummate gateway-to-the-Net digital format."

CDs, while digital, were very much a traditional physical medium. They were portable, but only in physical form. As they emerged, so did the personal computer and networked technologies which introduced not just digital processing but distance reduction into homes and radio stations. During the early 1980s, while it would have been possible to capture and even transmit digital audio data on a computer, the available storage, processing power and transmission speeds as well as the limited sound capacity of most early computers limited the scope of possibilities.

As networked technologies became popularized, communities in the United States and elsewhere began to share information through their computers. Before the widespread availability of the Internet after 1991, and for some time following, such exchange took place through what became known as electronic bulletin board systems or BBS. Limited exchanges of messages, software and images were the primary fare on these volunteer systems. Early online systems such as America Online and CompuServe allowed similar functionality for a fee.

As connectivity increased, greater connection speeds and computers with the ability to play sound became more popular. During the early development of several phases of the multimedia PC standards, uncompressed sampled PCM data as .wav files with relatively large storage footprints were common in representing actual sound, music or speech. For some time, these existed alongside less demanding (in terms of data storage and throughput) MIDI files that stored and reproduced music only by storing notes and durations and reproducing these on the host computer.

For all the swirling and sometimes competing technical developments, audio content on the Internet, outside of music and more in the mold of radio programming, began relatively early and sought to spread itself using whatever technical means were available. In 1993, only two years after the general public gained access to the World Wide Web, Carl Malamud launched an Internet audio effort called the Internet Talk Radio Service on March 31 through his nonprofit enterprise known as Internet Multicasting Service in Washington, DC. Sandberg (1994, p. B9) described the initial audio portion of Malamud's Internet Talk Radio service as "offering canned, 30-minute interviews called 'Geek of the Week,' which even less-powerful PCs with 'sound cards' can pluck from remote computers and download for a listen."

In an interview with *Broadcast and Cable* magazine in 1993, Malamud explained that "sound is just another kind of data . . . and most modern computers now have a speaker" (Viles, 1993, p. 27). However, despite the relatively primitive state of both the personal computing devices and the networks connecting them (primarily dial-up modems at the time), Malamud already had a vision for audio programming that well exceeded the technical achievement of accessing a sound file from a remote machine. Malamud's global view of radio on the Internet echoed in an interview with Thailand's *Bangkok Post*, as Waltham (1993, p. 1) reported:

> His ultimate objective is to stimulate "desktop broadcasting" and the aim of his programming, initially to be half an hour a week, is to serve as a world trade information service on networking in "this new global village."

Malamud pointed to the development of numerous options for audio transmission on the Internet by 1993 and the emerging possibilities of reception for even dial-up users at the time. However, like many others at the time, Malamud also saw the possibilities for fundamentally changing radio through the interactivity inherent in the new Internet environment. According to Waltham (1993, p. 3), Malamud hoped "to provide a truly active medium with active listener participation . . . combining the professional programming of radio with the worldwide reach of the Internet to create a brand-new medium."

Malamud (1993, p. 2), in promoting the service, emphasized the global scope as well as the ways in which this "radio" was at only a conceptual starting point, but also the launch pad for a "new medium":

> Radio is the initial metaphor, combining interviews, news, and analysis with professional announcers and musicians to form programs. . . . Radio is only a metaphor for this new medium. The user can treat the files as random-access radio, starting and stopping the programming, or changing the order of the show.

For Malamud (1993, p. 2), the new medium would be something "beyond radio" and like television's early likeness to radio, Internet radio would start as an analogy to its predecessor but quickly modify itself to suit the mew medium and its new possibilities:

> The obvious extension is to add data types, starting with graphic images and video streams. . . . New data types can be combined with new ways of linking the listener to the source of the programming. Interactive game shows and talk shows where the listener can double click to participate are two examples.

At the time, however, the technical challenges remained substantial even as Malamud cast his vision of global participation to include some 106 countries. Despite some development of streaming capacities later in its operations, Malamud's efforts were somewhat different from the eventual direct streaming from broadcaster to user that would evolve. Also, despite its relatively rudimentary nature, Malamud found himself, before too long, in conflict with US authorities such as the FCC, who were in charge of broadcasting and because of whom the Internet radio pioneer found himself emphasizing that his operations were non-commercial and provided content free of charge.

Malamud's radio (or radio-like) content could be found on the Web well before more advanced technologies were available and perhaps well before large audiences would have the chance to experience it. He closed his operations (which also envisioned television and live event streaming) in 1996, perhaps having made his point that the World Wide Web could act as a conduit for other media forms. Though Malamud, somewhat in advance of the rest of the industry, began to explore the possibilities of live transmissions, he achieved more recognition for pioneering digital audio content as a downloadable item, whose production as a digital file and distribution across the Internet through the World Wide Web was feasible.

As with many other technologies, developments in digital audio included a broad range of initiatives conditioned by existing hardware and science in the field. Efforts continued in terms of making audio files smaller and, related to this, being able to transfer them across networks. Thus, while some users and developers experimented with compression approaches that would decrease audio file sizes for users, others were approaching the problem from a different direction. Instead of simply trying to compress files, they were exploring the possibilities of feeding these files across the network progressively so listeners could begin to consume the audio before files were downloaded in their entirety, an approach that would become known as "streaming."

Companies such as Real Audio (established in 1994) with their Real-Player technology and Liquid Audio (established in 1996) for a while emerged as leaders in the realm of streaming audio. Competitors like Windows Media also offered possibilities. For the user, these competing and incompatible file formats required installation of separate client software. For producers of audio and streaming content, these competing formats required not only separate encoding, but often dictated investments in specialized servers that could deliver the audio streams. The quality of the streams that these early systems delivered suffered from limitations imposed on the state of the compression/decompression technology, the bandwidth available to both producers and consumers at the time and the computing resources at both ends. Progressive Networks RealPlayer compression/decompression (CODEC) scheme on its initial offering was characterized as being in the region of AM radio quality, while, by 1995, version 2.0 was being compared to FM radio quality ("The Enter*Active File," 1995).

Into this mix emerged a service known as Napster and a data compression system known as the MP3. The so-called "MP3" is a data compression technology that is properly known as the MPEG1-Layer 3 specification, arising out of the work of the International Standards Organization's (ISO) Motion Pictures Expert Group (MPEG) and contributions by private and public research interests including Thomson, Royal Philips Electronics, AT&T and the German Fraunhofer Society (Heingartner, 2007).

By mid-1998, outside of technical magazines or electronic bulletin board systems, reports in the mainstream media publications were touting the importance of (and exuberance over) the MP3. Blackwell (1998, p. K1), for example, wrote that:

> It's at the very bleeding edge of multimedia on the Web. MP3—short for MPEG 1 Layer 3—is a technology for making compressed, though still high-quality, digital recordings of music that can be stored on a computer hard disk and played back using shareware software on a multimedia PC or Mac. In short, it's taking the Net by storm—and causing a storm of copyright controversy in the music recording industry.

While the confluence of technological advancements including the MP3 made audio possible on the World Wide Web, it was the service called Napster that popularized the notion of sharing music files starting in 1999. The notion of digital audio seamlessly crossing distances became a reality.

As with other technologies, particular combinations of devices, users and various social factors drove development in myriad ways. Thus, beyond the clear appeals of portable digital music, there were other forces at work. MP3s and portable music players combined with the ability to download content (often for free) fueled not only an explosion in the digital consumption of music, but also, as some scholars have noted, in the production and consumption of audio program content (Swanson, 2010). One of the social drivers of such audio content was religion. Churches and religious leaders were quick to exploit the relatively cheap and easy route of making digital recordings of sermons and lectures and making those available for download to their followers—providing what some observers have termed "religion on demand" (Ralli, 2005).

These developments also brought into focus the potential uses of the new networked technologies and the prospects of digital audio beyond simple downloadable files. Even as early as 1998, media reports touted the existence of "an 'all request' MP3 radio station originating in Canada" that could be accessed through the World Wide Web (Blackwell, 1998, p. K1). In 1998, Ambrose wrote of listening to streaming radio from South Africa (while in the United States) that was already pushing the limits of what could (or should) be considered radio and demonstrating global ties as well:

> On a tip from a friend I logged into Qradio (www.qradio.net), and for the past 10 days l have been tuning in to live Zulu radio every time l sit down at my computer. Ukhozi is one of four South African radio stations that are streaming live over the Intemet, just one feature of the new, innovative Qradio. Underwritten by Quincy Jones, Qradio is currently bringing a huge resource online for people interested in South African music. In fact, it does much more: It brings South Africa itself online. (p. 32)

Such efforts were part of a growing set of initiatives from enthusiasts to harness the reach and power of the World Wide Web, particularly as it began to evolve from its textual roots into imagery and sound. Video, while also a popular aspiration, required greater advancements in bandwidth, computing power and compression technology before becoming a mainstay of the Web.

As the ability to deliver audio content in real time (or something close to it) eventually became a reality and despite the competing streaming platforms and formats, industry monitors became interested in tracking the growth of radio streaming operations available on the Internet. This tracking started as early as 1997, and by January 2000, the tracking numbers indicated something in excess of 3,000 radio stations streaming on the Internet. Approximately 48 percent of those operations were based in the United States and Canada where Internet access and broadband access were greatest. However, despite what was known as the digital divide that showed developing nations lagging in adoption of the digital global networked technologies, some 42 percent of the streaming operations were international stations moving their local programming online ("Radio Broadcasters," 2000, p. 31).

This high rate of adoption, however, belies something of the uncertainty of audio streaming during that period. The alphabet soup of compression schemes and formats was still in play including RealAudio, Streamworks, Internet Wave, TrueSpeech and ToolVox (Jacso, 1996) with debates raging about how to achieve the best results over the slowest connections and costs for implementation of streaming services ranging from free to the $2,500–$3,500 range to start and with prices escalating if the producer wanted to offer greater numbers of streams to listeners.

Pfaff and Toma (2003, pp. 302–303) described the various options open to broadcasters and would-be broadcasters at the time in trying to balance the need for compression with the need to maintain intelligibility in transmissions, identifying the major systems on offer at that time as including Microsoft Media 4, MP3, RealNetworks and Ogg Vorbis, which varied in their performance at different rates, the legal and financial burdens arising from their ownership and use and numerous technical choices having to do with different codecs and options. Investment, particularly with direct purchase of equipment and software, involved risks over which standard would eventually dominate and the chance of ending up with obsolete capacity (e.g., IBM's earlier ambitious but ultimately failed Bamba system for audio and video).

Adding to this confusion over formats and codecs, manufacturers were also introducing new devices predicated on the success of audio and radio streaming on the Internet. Palenchar (2000, p. 27), for example, wrote of tabletop Internet radios from dot.com startups AudioRamp.com and Kerbango that featured "built-in 56K modems and software that accesses streaming audio websites" without a PC. Palenchar (2000, p. 27) described the two offerings from AudioRamp priced at $399 and from Kerbango at under $300 with some attention to the alphabet soup of formats and codecs that plagued the consumer as well as the radio producer at the time:

> AudioRamp has developed the tabletop iRad-C AM/FM/CD radio and the iRad-C, an audio-component-style Internet-audio tuner that can be plugged into an A/V system. . . . Both AudioRamp products play back streaming audio in the Microsoft Media Player and Real Networks formats. They also download and store music files in the MP3, Microsoft Windows Media Audio (WMA) and Real Networks formats.

Meanwhile, personal computer users continued to be the most common recipients of audio streams. In 2000, computer enthusiasts, for whom downloading MP3s was already a "traditional" method of obtaining audio, were touting the "instant gratification" of audio streams "that begin playing seconds after you click a link" (Heid, 2000, p. 1). "Streaming audio's appeal is its immediacy," wrote Heid (2000, p. 1), who also touted the ability, with streaming audio, to "hear radio stations located across town or across the globe" or to listen to radio programming on your schedule.

The early popularity of audio streaming and its connections with radio became evident among personal computer users despite the proliferation of formats and standards (or, for more technical users, perhaps because of that added complexity). The fact that radio stations were already warmed up to the concept of simulcasting broadcast content to online streams, and an indication of Internet user interest, might be found in the development of listening software front ends such as Inklineglobal's WinFM, which the publisher described in 2001 as allowing you "to listen to streaming audio over your Internet connection," adding that "it comes with preloaded radio stations and you can add your favorite local radio station if they have streaming audio set up" (Inklineglobal, 2001).

By 2004, several software developers (including one hosting a site called stationplaylist.com) were also promoting their online streaming broadcast managers that would assist radio stations in creating and automating playlists for broadcast while offering the flexibility to employ the different streaming formats that were available at the time. Others, such as the DRS2006 radio automation system, relied solely on one format (in this case the MP3 variant MP3Pro) or another.

SEVEN

Local Radio and Global Audiences on the Internet

By the late 1990s and early 2000s, several technical factors were evolving in such a way as to provide radio practitioners and managers with new opportunities. Audio had already become a digital product, it had become manageable on the computer processors and storage media that were available and audio compression and streaming had evolved to a point where live audio could be provided to a geographically diverse pool of listeners once they were connected to the Internet. Within that pool of users, expanding both in the United States and abroad, users were also increasingly likely to be connected by broadband rather than dial-up connections (Mohammed, 2007). Numerous radio-like services such as Live365.com were also emerging on the World Wide Web. All these factors made migration of radio to the Internet more technically feasible.

While audio became more easily available online, various dot.com startups jostled for supremacy in various audio-intensive endeavors such as streaming music services. Battles raged over the legalities of music downloading, the potentials for monetizing music and other audio content online, the adoption of one or another streaming standard and the feasibility of various digital rights management systems. Despite (or perhaps because of) all these uncertainties, broadcasters around the world began to take notice of the potentials of radio on the Internet even as the emerging possibilities threatened to force a complete rethinking of what radio itself was and its place in society. Such concerns were exacerbated when, for example, digital jukeboxes like Live365.com gained popularity and listeners began to think of such services as Internet radio. Partly as a result of such trends, scholars like Tacchi (2000, p. 292) started asking direct questions regarding conceptualizations of radio past and present:

The move from broadcasting to the net raises the question of whether net.radio is radio. In order to investigate this it is necessary to consider what radio as such might be. Radio can be said to have certain characteristics, but the evidence suggests that radio is what history says it is: it has no essence since it has already taken, and continues to take, different forms. Radio is what it is at a given time, in a given context of use and meaningfulness.

Similarly Black (2001, p. 398) called the term "Internet radio" itself a "blatant and wishful . . . marriage of two terms" and asked (2001, p. 397):

> Why should an audio signal delivered through the Internet be called "radio" in the first place? . . . Do listeners to Internet audio streams count as radio listeners? Or is "Internet radio" a different medium from "radio"—and, if so, why has it borrowed the name?

This confusion and redefinition would, rather than be resolved, continue into the development of Internet radio and its evolution into diverse forms as Lasar (2016, p. xii) noted:

> The Internet radio era redefined the very concept of "radio." Rather than being understood as AM and then FM, it now became a vast hodgepodge of technologies, applications, social practices, and things. Audio files posted on Web pages became radio. Services such as Pandora became radio . . . "podcasts," became radio. Playlists on services like Spotify and YouTube became radio. Music shared by participants on chat rooms became radio. The term "radio" gained many new technological allies and friends, but lost the clarity that it enjoyed in the twentieth century.

Radio as a conventional broadcast undertaking was being called on to compete with new conceptualizations of radio as the digital networked technologies emerged into mass adoption and to question some of the long-running assumptions about "traditional" radio itself. Among these assumptions were technical ones such as the general understanding of radio as a one-way broadcast medium which it evolved to be despite the two-way nature of its predecessors such as the wireless. The interactive and two-way nature of the Internet and Web-based technologies prompted theorists and practitioners to deal with not just the prospect of enhanced feedback to broadcasting, but the possibilities of direct two-way communication options.

As with traditional airwave broadcast radio, the notion of funding also quickly became an issue which broadcasters had to address. The possibilities of advertising on Internet audio or radio services became entwined with broader arguments (which started soon after the Internet began to gain public popularity) about whether it was appropriate to include advertising content (or indeed any business activity) on the network. Purists saw the public, nonprofit, collaborative and international nature of the Internet as anathema to advertising. Investors and innovators saw it as a frontier in which claims were to be staked. By the early 2000s, as radio began to stake its claims to the digital frontier, so many businesses were starting up and failing online that the question of whether advertising was appropriate on the Internet became little more than an anachronistic debate.

In addition to the difficult prospects of securing revenue streams, the costs of going online, whether as a purpose-built Internet radio operation or as a streaming version of an existing real-world station, were often substantial at first. Whether a station invested in streaming servers or (as eventually became more common) outsourced the streaming side of the operation to companies dedicated to such services, there were significant costs to broadcasters, and major players such as NetRadio.com (an Internet-only service that offered multiple streams using RealPlayer) were among the early few who were able to realize their streaming aspirations.

However, from within the array of competing standards and formats and with Internet access becoming more prevalent throughout the United States and abroad, numerous less powerful audio broadcasters began to venture onto the Internet as well, reaching potential audiences for far less money and effort than a traditional radio broadcasting operation would demand. By 2003, Ting and Wildman (2003, p. 31) estimated the cost of establishing an online radio station in the United States at some $8,700, while a traditional terrestrial radio broadcasting operation might typically cost around $280,000. As the costs and challenges of streaming audio diminished, more amateurs could be found doing radio or radio-like things online. Smethers (2016, p. 21), for example, noted the activity of "thousands of artists, hobbyists, and potential entrepreneurs" who have used audio streaming on the Internet "as a low-cost way to 'broadcast' or to at least share their renditions of music, talk, or other artful fare" and argued that this phenomenon prompted notions of "the democratization of radio" in the early 2000s.

On the content side, broadcasters both real and virtual had to face multiple legal and financial issues surrounding payments of royalties for Internet broadcasts. Music royalty payments were at the heart of legal battles between the Internet broadcasters and artist groups and record labels in the United States. Artists groups and labels clamored for revenues they felt they were owed from their material being broadcast online. Internet radio services, while acknowledging the need for such payments, were opposed to the music industry groups' attempts to modify their existing systems of royalties used for radio stations. Internet broadcasters felt that conditions imposed under the Digital Millennium Copyright Act favoring the music industry unfairly burdened the Internet operators whose online revenues were quite small.

When the market failed to yield a solution to this impasse, the US government stepped in through the US Copyright Office under the Library of Congress in June 2002 to propose a compensation rate of 7 cents per song per listener (down from its own committee recommendation of 14 cents per song per listener). This approach was not popular with either side. As the debate and jostling for positions continued, the US Congress was repeatedly involved in sorting out the contentious situation. A 2004 congressional report entitled *Internet Streaming of Radio Broadcasts: Balancing the Interests of Sound Recording Copyright Owners with Those of Broadcasters* (House of Representatives Committee on the Judiciary, p. 2) described the two sides of the debate. Broadcasters felt that the Digital Performance Right in Sound Recordings Act (DPRA) of 1995 and the Digital Millennium Copyright Act of 1998 unfairly placed "new and unreasonable burdens on radio broadcasters," asserting that those burdens had "resulted in more than 1,000 US radio stations ceasing their Internet broadcasting operations," arguing for relaxation of what they felt to be onerous and damaging legal requirements. The RIAA, arguing for artists and record labels, sought to maintain the legal provisions because new and emerging technologies were seen to pose "a grave threat not only to the livelihood of music artists, but also to the advertiser-supported business model of radio broadcasters" (House of Representatives Committee on the Judiciary, p. 2). Negotiations thus framed were to continue for many years into the early part of the twenty-first century.

THE INTERNET, RADIO AND "HOME-MAKING"

Despite this and other hurdles, the presence of radio and radio-like content on the Internet was spreading, not just in the United States but also across the globe wherever the conditions for Internet access made the World Wide Web accessible to users. In 1999, a newspaper report from Melbourne, Australia (Schulze, 1999, p. 1), touted the projection that "Net radio" was tipped to explode with the number of listeners (though small at the time) "expected to expand fast." An AC Nielsen study estimated Australian Internet radio users at about 23 percent of Internet users in that country in 1999 but pointed out that Internet radio listeners in the United States had jumped from 18 percent to 30 percent of Internet users during the period 1998–1999 (Schulze, 1999, p. 1). Among the reasons for listening among Australian Internet users were simple curiosity, as well as the intrigue of being able to receive "overseas or interstate stations" (Schulze, 1999, p. 1), but local and national radio operators were already showing interest, including plans for Internet-only stations:

> Australian companies are already testing the Internet-radio sphere, with Telstra last week announcing an exclusive agreement with the Internet radio company Digital One to launch five Internet radio channels this month and another five next year. Australia's largest radio group, Austereo, has created the first of what it says will be many Internet radio stations, ozchannel.com.au.

Industry analysts were beginning to face the emergence of these technologies that, despite their infancy, were signaling immense (even "revolutionary") changes to come, including global reach, as Alderton et al. (pp. 126–127) noted in 1999:

> Today's fledgling Internet radio forces listeners to sit at their computers and offers tinny sound no better than they could get from a cheap AM receiver. Yet these drawbacks will soon vanish: wireless Internet access will bring the Walkman's freedom of movement to Internet radio. Besides, it offers something no other radio medium provides: worldwide coverage.

Coyle (2000) categorized such early radio forays on the Internet into three categories: 1) extensions of traditional broadcast station content into online streams, 2) on-demand radio or access to pre-recorded content online and 3) interactive content including audio, visual and textual material that might, today, also be termed social.

The motivation to migrate or supplement traditional over-the-air broadcasts with Internet streaming was particularly attractive to small stations with limited terrestrial signal reach and who sought to expand their audiences beyond that reach. Ogun Radio in Nigeria, for example, was established in Abeokuta in 1976 with a mandate to serve audiences in Ogun State but with coverage that would eventually include the neighboring states of Oyo, Lagos, Osun, Ondo and Edo as well as parts of the neighboring country of Benin ("Ogun Radio: About Us," 2018). However, it has established both streaming services and social media presences that include global audiences and feedback from listeners outside its coverage area in Nigeria and those living abroad.

These once purely local media have often formed part of the set of what Bonini (2011, p. 869) has called "home-making" tools in the lives of those separated from their homelands. Bonini (2011, p. 879) described the various methods that a Filipino migrant in Italy used to virtually engage with "home" that included "a streaming of the Philippines' public radio" playing on his computer at all times. Hiller and Franz (2004) earlier chronicled the efforts of diasporic members of Canada's Newfoundland community to create a virtual sense of home by listening to streaming versions of online radio from Newfoundland. Ogunyemi (2006) described the appeal of African radio streams to the African diaspora in the United Kingdom including the use of traditional music and African chart music to both gain audiences and to encourage interaction with their websites.

In the oil-rich Arab Gulf states, expatriate worker populations from numerous countries often rival the natives in total numbers and include manual laborers as well as bankers, engineers and other professionals, many from the Indian subcontinent. Abraham (2012, p. 131) has described the emergence of Indian ethnic media in these states, broadcasting terrestrially in the diaspora country but often using the language of the audience's home country. India's Radio Mirchi, for example, relays popular programming from India and presents this and other content on terrestrial broadcasts and Internet streams hosted locally in the United Arab Emirates.

In addition to traditional radio stations migrating online with new-found international reach, Internet-only, radio-like audio services also emerged to serve far-flung diasporic audiences. The Radio Africa Online service, for example, is primarily a rotating jukebox of African and Caribbean music that started in January of 2002 ("Frequently Asked Questions," 2018) but also features a podcast production, song request service, a discussion forum and live events. Starting as a Congolese music service, the web service quickly evolved into a radio-like portal spanning a Pan-African and African diaspora audience with English, French and traditional African languages featured in its broadcast streams and other content. In these diverse examples of Internet radio efforts, we may also find an acute awareness of the changing nature of audience engagements to which Coyle (2000) had earlier alluded, including a greater diversity of media options including not just home computers but also mobile phone apps and the increasing abilities of audiences to move beyond passive listenership into active involvement with the stations and other audience members to form virtual communities at home and in the diaspora (Mohammed, 2017). Further, several studies (Kuhn, 2011; Mohammed, 2012; Mohammed & Thombre, 2017) have suggested that despite the fact that many local stations have deliberately fostered international reach and are acutely aware of their international audiences, they often tend to display what Kuhn (2011, p. 44) has called a "local inclination" manifest in "the pervasiveness of local issues, which become relevant to people living in other countries."

ETHNIC STATIONS IN THE UNITED STATES

Among the stations that were most driven to make the move to the Internet were those for whom expanded reach was not something they could achieve given the small size of their local target markets; among these were small broadcasters in many different countries with potential international audiences and so-called "ethnic" stations with widely dispersed audience members.

The present work cannot sufficiently address the problems associated with the term "ethnic," its connotations and its connections with the fallacious concept of race. For our present purposes, we will simply acknowledge that the term "ethnic media" has been used in a broad sense to denote non-white, non-mainstream media in the United States. However, in other contexts, the same or similar terms may denote media that focuses on one particular cultural heritage as opposed to others with varying peculiarities and boundaries that may serve to mark the "ethnic" against the "mainstream" (Bratu, 2014; Yu, 2015).

Ethnic broadcasting did not emerge in a vacuum but from a tradition of minority and immigrant press in the United States that captured the attention of sociology pioneer Robert E. Park, whose description (1922, pp. 71–72) of that tradition was less than laudatory:

> The peasants are sentimental; the editor prints poetry for them in the vernacular. He fills the paper with cheap fiction and writes loud-sounding editorials, double-leaded, so that they will be easily read. The readers are very little interested in abstract discussion, so the paper is more and more devoted to the dramatic aspects of the news and those close to their own lives, the police news, labor news, and local gossip. Sometimes the publisher is himself an ignorant man, or at least not an intellectual. . . . It is said that one of the most successful Chinese editors in America cannot read the editorials in his own paper because he does not understand the literary language.

Ethnic broadcasters in the United States can trace their roots back to the early multi-language broadcasts that were featured on many stations in cities with immigrant communities who still spoke their native tongues. As we have seen in previous chapters, the coming of World War II and suspicions about what might pass unnoticed in German or Italian had a dampening effect on these broadcasts. However, that is not to say that broadcasting aimed at minority groups disappeared. In fact, such broadcasting took on a variety of forms. In many cases, the broadcasts would be featured in limited slots on otherwise mainstream stations. WSBC in Chicago was one such station which featured broadcasting aimed at African-American audiences starting in November 1929 with a program entitled *The All Negro Hour*. Such programming was exceptional at a time when predominantly negative stereotypes of Africa-Americans could be found in mainstream offerings such as *Amos and Andy* and *Beulah*. Isaksen (2012, p. 755) has described *Amos and Andy* in the following terms:

> Debuting on NBC in Chicago in 1929, the 15-minute six-night-a-week show captured 53 percent of the listening audience—about 40 million people—thus, profoundly re-inscribing the legacy of blackface racial meaning in America. Two white vaudevillian actors, Freeman Gosden and Charles Correll, fell back on the seasoned showbiz formula of blackface minstrelsy, but in this technological version, the blackface was aural not visual; using racial ventriloquism, white men put into the mouths of blacks a racial representation that was highly distorted, falsified, and inauthentic.

The Chicago Defender, an African-American-owned and -oriented newspaper noted, in commenting on a December broadcast of WSBC's *The All*

Negro Hour, that "these all Negro programs" were "becoming very popular" at the time ("Broadcasting," 1929, p. 7). Soon after the launch of *The All Negro Hour,* newspapers reported that the WSBC station management had hired the African-American producer of *The All Negro Hour,* Jack L. Cooper, as Chicago's first official African-American announcer ("Jack Cooper Is Announcer," 1930; "Jack Cooper Proves," 1930).

As early as 1938, NBC introduced network broadcasts of something called the *National Negro Radio Hour* originating from Cleveland, Ohio. The *Atlanta Daily World* (also itself an "ethnic" newspaper) called the first broadcast of the *National Negro Radio Hour* "something to be proud of" ("National Negro Radio," 1938, p. 6). *Variety* magazine ("Radio-Television," 1966, p. 56) reported that the Radio Advertising Bureau in October of 1966 had established a Spanish-language division due to the "growth of ethnic radio."

Neigh (2013, p. 269) noted that for some, like poet and journalist Langston Hughes, American radio represented a bastion of racial privilege well into the 1940s, being "used by white-dominated America to maintain its power" and remaining inaccessible to persons of color except on terms and in contexts dictated by the dominant class station owners. This situation would change (but not completely) with the emergence of minority programming and minority station ownership.

Kahlenberg (1966) traced the start of African-American radio stations to 1947 with the single station WDIA in Memphis, Tennessee, and with some 270 stations with varying amounts of content for African-American audiences in 1954 and over 500 in 1966. In 1951, excitement over so-called "Negro Radio" was manifest in efforts at national distribution and news of plans for an *All-Negro Radio Show* to feature "top names in the rhythm and blues field" was sufficiently important to make the front page of the *Atlanta Daily World* newspaper ("Launch All Negro," 1951).

African-American audiences were only one among a range of target markets being explored in the 1960s as a 1962 estimate cited more than 900 radio outlets broadcasting in more than 40 languages and dialects "ranging from Spanish and Italian to Hindustani, Pennsylvania Dutch and Croatian" ("Ethnic Group Programming," 1962, p. 42). This, as television, with its greater production complexities, greater costs and higher economic risks, found difficulty in addressing ethnic audiences.

In 1969, *Broadcasting* magazine announced the launch of the Soul News Network set for March 7 of that year (1969, p. 9). The network, originating in Washington, was "designed to provide audio feeds to ethnic radio stations" and would "concentrate on news and personalities of interest to minority groups" ("At Deadline: Ethnic News," 1969, p. 9). This network provided content to major ethnic broadcasters of the time that included WLIB, New York; WAOK, Atlanta; WJLB, Detroit; WAWA, Milwaukee; WEUP, Huntsville, Alabama; KNOK, Fort Worth and KAPE, San Antonio.

Around the same time, some controversy surrounded African-American-oriented radio stations in the United States. This was particularly due to the fact that in 1969 *National Review* publisher William F. Buckley (who had argued in his publication that civil rights should be opposed) bought his way into three black-oriented stations in New Orleans, Memphis and Houston (Ferretti, 1970). The motivation for such a move was probably the fact that African-American-oriented stations, numbering about 310 nationwide at the time, were estimated to be earning some $35 million in advertising billings for their owners, the vast majority (95 percent) of whom were white (Ferretti, 1970). Buckley's involvement revealed the reality that most of the substantial revenue being generated was enriching persons outside of (and even opposed to) the community to which the broadcasts were aimed.

African-American radio and other ethnic stations continued to operate in the United States, but with a history of economic challenges related to their small audiences and limited signal reach. The economic realities of operating with small audiences and limited budgets could be seen in the ebb and flow of station numbers as ethnic stations often failed to stay on the air. Smith and Cornette (1998a, 1998b), for example, estimated that after the start of Native American radio stations in 1971, approximately 35 Native American stations were in operation across the United States by 1998 (up from 11 in 1980). However, the number of terrestrial Native American radio stations had slipped back to about 24 in 2004 and then rose to about 50 in 2013 (Dukepoo, 2013) as digital and streaming technologies changed and became more accessible.

For many of these Native American stations, their primary objectives had more to do with serving the linguistic and cultural needs of their audiences than with reaping profits. Their operations often depended on the input of local volunteers and their services to the community included numerous activities in which other more mainstream commercial stations were not likely to engage. Thus, in addition to covering government meetings at Window Rock, Navajo Nation station KTNN was also known for its community announcements and services. Staff members at the station (in 1996) were fond of telling visitors the story of two children who were found at a Window Rock shopping center after their extended family had driven off and forgotten them behind. In a time before cellphones were widely available on the reservation, and knowing that virtually everyone was tuned to the local station, the staff broadcast the names of the children and the fact that they were waiting to be picked up. Before long, the family heard the message and returned for the siblings.

KTNN was also somewhat famous as a cultural broker of sorts. In 1996, it became the first Native American broadcaster to cover the NFL Superbowl in the local *Diné* language to an audience of some 250,000 people on January 28 (Odeven, 2005). In 1997, KTNN also caused something of a sensation when they hosted Rex Redhair, also known as "the

Navajo Elvis." Hudis (1997, p. 47) reported that Redhair (a local juvenile probation officer and social worker) was actually "a crooner with a heart of white polyester" who about two years earlier had demonstrated his talents to the KTNN staff:

> Redhair walked into the KTNN studios and started recording songs, Elvis-style, which the station broadcasts to the Nation. His big hit is "Don't Be Cruel."

After rising to fame, Navajo Elvis gave a live performance on the air, which, according to Hudis (1997, p. 47), "almost caused a stampede":

> "All these women, 40, 50 and 60 years old, left work and crowded down here," says KTNN program director Scott Scarborough. "He was all dressed up like Elvis and signed autographs and kissed the women," adds Jenna Yazzie, KTNN's midday on-air personality.

Radio has also played an important role among other native peoples such as those in Australia where indigenous-produced radio programming has been an important feature of community radio since the 1970s (Forde, Meadows & Foxwell, 2009).

At the turn of the twenty-first century, Venugopal (2001) described the efforts of several broadcasters in the United States who featured material catering to diasporic Indian populations with their roots in the Indian subcontinent. Venugopal (2001, p. 20) characterized these broadcasts as "weekly rituals for hundreds of thousands of South Asian immigrants: the soundtrack to the desi experience," some of which provided "good, solid Indian music and Indian values . . . the kind you'd raise your kids on." These broadcasts included programs such as a weekly *Music of India* broadcast in Fairbanks, Alaska, where less than 1,000 families of Indian descent were thought to live, Anil Srivatsa's *"Anil-Ki-Awaaz"* weekly broadcast carried on several stations in Washington, DC, and California as well as weekly programs in Florida, Houston and Dallas. Venugopal (2001, p. 20) traced these efforts back to 1975 when Brij Lal (a former All India Radio and Voice of America correspondent) started an Indian-themed radio show at Fordham University in New York, which would become "the standard for later Indian radio shows":

> Initially, it was only in the large urban markets, where the first waves of immigrants congregated, that such programs could be heard. But as the diaspora made its way into smaller pockets of America, college towns and semi-rural enclaves, the medium followed, driven less by commercial incentive than by the hope of community sustenance.

Not long after Brij Lal started his program in New York, Chicago-based diaspora Indian communities were also developing their own programming. As Rangaswamy (2000, p. 322) described it in 2000, ethnic radio programs, a popular source of entertainment for South-Asian Indians in Chicago, started with informal efforts in 1976 and continued into the twenty-first century:

> *Jhankar* is an AM/FM radio show that broadcasts over a seventy-mile radius around Chicago from recording studios in Bloomingdale. Started by an electrical engineer as a hobby in 1976, the show plays popular Hindi film songs from the "good old days" of the 1950s, 1960s, and 1970s, an era which the pioneering Indian immigrants remember with nostaligia. Other AM radio shows that broadcast popular Hindi music are *Sangeeta, Tarana* and *Naubahar.*

Yet despite their popularity and usefulness to their communities, ethnic radio stations and occasional programs tended to encounter financial problems, particularly as conglomerates began to buy up local radio stations and depress the job and advertising markets in the process. Alexander (2008, p. 106) drew on the connection between the spread of conglomerate ownership with its focus on the national market and the threats to survival for local ethnic media:

> To reach this coveted market, large broadcasters use nationally syndicated radio shows in an attempt to expand audience reach and market share and to increase revenues. As a result, local programming, once an indispensable staple of black radio, is quickly becoming a distant memory. And since syndicated radio provides a platform for national advertising, the stations with local programming can only compete for local advertising.

Even some of the more popular ethnic stations such as New York's legendary Caribbean diaspora broadcaster WLIB found themselves the victims of market forces and disappeared from the airwaves to give way to alternative formats. Their struggles would be substantial over the years, particularly with the high overheads and operating costs associated with traditional analog radio production and transmission in which their listening audiences were limited by their signal reach. Attempts at radio networks, including the somewhat successful American Urban Radio Networks that survived for some three decades (Baraka, 2001), were the primary means by which small ethnic stations could reach larger audiences.

STREAMING AND ETHNIC RADIO

Johnson (2005, p. 26) has chronicled the survival of a small ethnic radio station, KJLH, in Los Angeles, California, whose strategies included using the Internet to expand their reach well beyond what might be possible with their terrestrial transmitters:

> It is the only independently and locally black-owned and operated commercial radio station in Los Angeles, and its coverage area is limit-

ed by its 5,600 watts (originally 3,000 in 1992). But the Internet has connected its South Central listeners to others across the USA. Emails and calls every weekday morning come into the studio from listeners on the East Coast and in Mid-America who have tuned into the call-in program *Front Page*.

Among the key drivers of KJLH's migration to the Internet were the national attention that its *Front Page* program attracted since the 1992 civil unrest in Los Angeles and the availability of commercial streaming services. These third-party providers, such as BroadcastUrban.com, reduced the uncertainties and overheads of investing in streaming servers and associated technologies that radio stations would have previously had to contemplate to achieve an Internet presence.

By the year 2000, *Music Business international* ("Radio Broadcasters," 2000) was commenting that all radio stations in the United States were facing the decision of whether or not to offer some version of streaming radio with some eagerly embracing the new realities while others hesitated due to concerns about cost and technological uncertainties. Small ethnic stations with limited terrestrial reach had the most to gain from this migration but were often least able to risk the investments to achieve it.

The quest among minority-oriented stations for national audiences was one which was being addressed by attempts for black-oriented radio networks dating back to 1954 and with several attempts along the way of varying success. With Internet streaming, even small stations could reach national audiences in a manner that was previously impossible.

In the United States, in addition to increasing the reach of traditional radio stations, streaming media also began to expand the (often global) possibilities of the ethnic media market. *India Currents* reported on the establishment of a service in San Jose, California, calling itself Homeland Networks that promised relayed media streams of content aimed at Indian immigrants delivered over the Internet branded with labels such as ROI (Radio of India) and TVOI (Television of India):

> It is a great time to be an India Immigrant in the US. Although half a world away, never before has he been closer to his roots. Telecommunications and the Internet have amazingly shrunk his world. . . . Even though we learn to live around it, there are sounds and sights from the native land that we crave, like that soulful Tamil melody or *bhajan* heard on the radio. ("Bringing India to You," 2000, p. 24)

Native American radio interests in the United States combined streaming technologies with satellite technologies to expand the possibilities of their broadcasts as well. The American Indian Radio on Satellite (now AIROS Native Radio Network) out of Lincoln, Nebraska, distributed programming through satellite and, later, over the Internet to affiliates throughout the United States starting in 1992. Both independently and as participants in networks, Native American, Canadian First Peoples and other indigenous broadcasters around the world have utilized streaming radio to enable their broadcasts to reach not only their contiguous communities but also their increasingly widely scattered diaspora communities and other like-minded groups internationally. Native Voice One Native American Radio Network, for example, listed participating and contributing stations not only in the United States, but also in Canada and Australia (Koahnic Broadcast Corporation, 2016).

In areas of the United States with Chinese diaspora populations, radio has developed catering to these ethnic and linguistic communities and streaming has added a direct international component to their content. KAZN AM1300, for example, is one of several Chinese stations operating in California. Established in 1993, the station broadcasts in Chinese languages including Mandarin on its terrestrial transmitter and repeats to an Internet stream boasting several hundred thousand listeners. KAZN started around the time that Chinese audiences in the United States were being described as "practically anonymous to the broader American radio market" (Knopper, 1996, p. 77). KAZN also highlights its role in the diasporic and immigrant experiences, noting its importance to the Chinese community in Southern California and to new immigrants to the United States ("About Us: KAZN," 2019). Several other California-based Chinese language stations have evolved since the 1990s including not just music stations but also operations such as the news-oriented *Sing Tao* Chinese Radio broadcasting in Cantonese and Mandarin on 1400 AM and 96.1 FM and streaming live in both languages (Sing Tao Chinese Radio, 2019).

Radio audiences for Chinese content are also able to access streams from numerous Chinese broadcasters including China Radio International and public broadcaster Radio Television Hong Kong as well as popular Chinese music stations such as Easy FM from Shanghai and Beijing and Love Radio broadcasting on 103.7 FM in Shanghai and streaming online. These stations' streaming operations also include social media presences such as on Facebook and Weibo, enabling and encouraging listener feedback from their local and global listeners.

Elsewhere in the world, some ethnic stations were also developing global aspirations. As media markets in developing nations liberalized and more radio stations evolved, broadcasters in those nations began to explore the prospects of expanding their reach and their markets not just nationally but, in some instances, also globally.

Indian Radio in Trinidad and Tobago

The small Caribbean two-island nation of the Republic of Trinidad and Tobago counts as slightly more than half of its population of some 1.3 million (The World Bank Group, 2010), the descendants of Indian indentured laborers who were brought to Trinidad during the 1800s and the early part of the 1900s to work on sugar plantations. Though there were opportunities for return to India after the term of indenture, the vast majority of these indentured laborers remained in Trinidad, often induced by offers of land or money to do so (National Council of Indian Culture, 2015; Lal, 1998; Northrup, 1995; Roopnarine, 2006). Debates continue about the nature of the cultural relationships and links between the transplanted and generationally removed diaspora in Trinidad with the Indian ancestral home (Samaroo, 1987). Despite an almost total loss of their ancestral languages, the Trinidad diaspora Indians continue to harbor (in varying degrees and in eclectic fashion) sentimental ties to India reflected in food, music, religious beliefs and other facets of life. Among the cultural ties (whether real or imagined) that are perceived is one of the consumption of Indian music—even though as far back as the early 1980s less than 3 percent of these descendants understood anything of any Indian language (Jayaram, 2000).

In and around 1947, when Trinidad and Tobago launched its first radio station, the linguistic mix of the indentured laborers and their immediate descendants was quite substantially more varied. Indeed, the launch of the first "ethnic" programming on Trinidadian radio was linked in some part to the fact that one of the figures invited to participate in the launch, Kamaluddin Mohammed, was not only a religious leader but also fluent in Hindi, Urdu, English and other languages including Persian and Arabic (Ghany, 1996).

When Mohammed embarked on the first ethnic broadcasts on Radio Trinidad (operating on 730 KHz AM), it was part of a broader context of ethnic broadcasting that already existed in the region. As we have seen above, documentary evidence clearly demonstrates that Kamaluddin Mohammed was listening to broadcasts from Guyana's pioneer of ethnic broadcasting, Mohammad Akbar, who was also broadcasting to a similarly displaced and constituted ethnic minority. It was also common in the Caribbean and South America for these diasporic Indians and their descendants to listen to shortwave broadcasts from All India Radio, where popular film music dominated for several years until 1947 when Indian post-independence policy deemed such content not sufficiently reflective of Indian national culture.

Indian ethnic radio broadcasts in both Trinidad and Tobago and in Guyana continued for many years on their local radio stations with limited offerings typically of a half-hour, once-weekly program, sometimes extended to an hour depending on advertising and song request revenues. In Trinidad this broadcasting niche evolved with live performances at first, featuring local bands. Traditional songs quickly gave way to an emphasis on Indian film music and bands eventually gave way to recorded media. However, the global context of transplanting Indian content to Trinidadian (and Guyanese) airwaves continued. Radio Trinidad exercised a monopoly on that content in Trinidad, as limited as it was and Guyana with its lone radio station experienced the same situation.

After a focus on nationalization of key industries in the late 1960s and 1970s, the government of Trinidad and Tobago nationalized one of the two radio stations in existence. The National Broadcasting Service (NBS) radio station (also known as Radio Guardian) operating on 610 KHz came under government control and a young broadcaster convinced programming leadership to allow him to produce a program of Indian music. These productions also took the form of weekly broadcasts of a half hour or one hour but were eventually supplemented by shorter daily programs of fifteen minutes' duration (Hanoomansingh, 2015).

Such was the state of Indian ethnic radio broadcasts in Trinidad and Tobago until 1993. Under the pressure of rising costs and public demands for a mass media environment less subject to government policy, the government had been exploring, for some years, the implications of granting more radio and television licenses to private operators. An attempted coup in July of 1990 by a group called the *Jamaat Al-Muslimeen* brought into sharp focus the need for a more diverse mass media system. Highlighting this need, in particular, was the fact that the insurgents, as a part of their attack on the government, took control of the lone television station and one of the two radio operations in an effort to convince the public that the coup leaders had seized power. Government attempts to jam their signals succeeded in part, but their attempts to reach the public by alternative frequencies largely fell flat outside of the capital (Searle, 1991).

With the dangers of a limited set of information sources fresh in the national memory, the government acted to divest national media and to allow additional private broadcasters to enter the market. When licenses were made available in 1993, a local advertising salesman proposed, with a small group of friends, to launch a radio station that would cater to the Indian ethnic niche audience previously served for more than four decades by occasional programs on weekends and short segments during the week. The idea found little traction among existing broadcast operations, the traditional wisdom being that such limited audiences would yield limited revenues. The small group of radio and advertising practitioners then decided to embark on the project independently as Radio 103 FM on 103 MHz and started not just a successful company, but an entire niche industry within Trinidadian radio, spawning imitators within a few years and counting the launch of some ten competitors in their more than twenty years of broadcasting (Henderson, 2015).

As 103 FM and other Indian radio stations in Trinidad began to gain popularity, two factors began to push them toward a global scope for their radio operations. One of these factors was the existence of a large secondary diaspora of Indo-Trinidadians, the descendants of the indentured laborers from India, who had migrated over the years to metropolitan centers in North America. The other factor that served to push these stations toward a global outlook was the deeply ingrained tradition of song requests and dedications on Indian radio content in Trinidad and Tobago.

Initially, song requests were a source of income that supplemented the advertising revenue that the host of a program on the radio would solicit and air. Kamaluddin Mohammed, in fact, paid his mentor, Mohammad Akbar, for song requests in the early days of their correspondences (Akbar, 1946). Later, the practice of charging for song requests faded away, but the thrill of writing in to ask for a selection to be played for a family member's birthday or some such event and hearing the dedication read on the air was a major draw for listenership that evolved over the decades of limited ethnic Indian content in Trinidad and Tobago.

When dedicated ethnic Indian stations began to gain popularity in Trinidad and Tobago, their use as a request and dedication service was a major component of their audience draw. Whereas the traditional route of request was the written letter, by the advent of Indian radio in Trinidad, landline telephones were the preferred medium—providing, in many cases, announcer feedback on when the request would air. However, the stations were barely able to reach all ends of the small, two-island republic, so that dedications and song requests for family members abroad in New York or Toronto (or even Miami) seemed somehow pointless with the relatively weak and low-range FM transmissions of these stations. This fact did not, however, stop audience members from making such requests.

In most cases, a request to a foreign relative would just be spilled into the ether, family members reveling in the mention of their names or their relatives' names on the air and telling their family members about the event. More enterprising listeners would record the reading of the dedication and play it back for their relative. Those who could afford the international phone bill would call up their relatives abroad and listen together until the request was announced (Henderson, 2015).

Trinidad and Tobago's ethnic Indian radio stations, recognizing the global dimension of their appeal, began to innovate to address audience members' connections with their families abroad. Among these innovations were occasional live station hookups between local Indo-Trinidadian stations and ethnic broadcasters in the United States.

The ethnic media situation in New York City, in particular, presents a plethora of diverse radio offerings at any one time. As we have seen, however, these offerings can be mercurial. Programming, and even stations, do tend to flourish and fade with time in this market. Given competition for the urban airwaves, however, in most cases, failed stations are quickly replaced by some alternative content operating on the same frequency. Ethnic programming catering to the Indo-Trinidadian and Indo-Guyanese diaspora markets living in New York City's neighborhoods has been a feature of the New York media-scape for many years (Tanikella, 2009) through a combination of dedicated stations and occasional program slots.

The mix of emerging Indo-Trinidadian stations in Trinidad and existing ethnic broadcasters in the United States aimed at Indo-Trinidadian diaspora communities provided fertile ground for the emergence of international station link ups. These took different forms, but generally featured disc jockeys connected by telephone to each other's studios in Trinidad and in New York. Listeners in either location could call their local stations and send greetings or song requests to their family members abroad. Restricted to particular programming segments, these special events would be limited to weekends when ethnic programming was usually run on the New York stations and when high listenership could be anticipated in Trinidad—important at that time to cover the relatively high expenses of international calling.

This particular form of the global ethnic radio connection was, however, only a precursor to the emergence of Internet streaming. The parochial Indo-Trinidadian stations, programming as they were for a specialized audience in a small country, found that they could increase their listener base and advertising reach by utilizing the global digital networks and the ability to stream audio, which was evolving and becoming accessible, even in a developing nation like Trinidad and Tobago by the early 2000s. From early experiments in streaming utilizing a variety of platforms, the Indo-Trinidadian radio broadcasters have evolved into sophisticated media streamers, feeding their fare of Hindi and Trinidadian music to their audiences in the North American metropolises and even further afield. Their streams include not only audio feeds but frequently, as well, video feeds from their studios and integration with social media and mobile applications. While listeners do still call the numerous (at various times more than a dozen) stations to send greetings and song requests, most of these are now enacted through Facebook posts and text (SMS) messages or through mobile apps within a larger context of interactions and parasocial relationships (Horton & Wohl, 1956; Papa et al., 2000; Schiappa, Gregg & Hewes, 2005) among listeners and station personalities (Mohammed, 2017; Mohammed & Thombre, 2017).

Content analysis of the Facebook pages of some of these stations indicate listenership throughout the Caribbean region, in the North American centers of New York, Miami and Toronto and even further afield in places like the United Arab Emirates (Mohammed & Thombre, 2017). Despite this global reach, however, listeners and station managers alike indicated that their primary focus remained the local station and domestic events (a sentiment echoed during interviews with station management at NCN communications in Guyana). The irony of the global influences that dictate the existence of Indian-oriented stations in Trinidad does not generally impress or bother the programmers, audiences or on-air personalities.

EIGHT

Global Dimensions of Hate on the Radio

The evolution of propaganda broadcasts from political mouthpieces promoting particular causes into tools of division and, in some cases, hate and even genocide was a gradual process. The calculated efforts of European nations to influence both friend and enemy in the geopolitical processes of (particularly) World War II ranged from relatively benign defenses of political activities (such as Mussolini's 1931 broadcast reassuring American audiences that there was nothing to fear from Italian Fascism) to virulent and racist propaganda attacks such as those the Nazi party launched against Jewish populations in Germany and elsewhere. The Nazi minister of propaganda, Joseph Goebbels, stated at the opening of the tenth Radio Exhibition of August 1933 in Berlin that: "It would not have been possible for us to take power or to use it in the ways we have without the radio" (1933, p. 198). In that same year, Nazi authorities used the radio to call for boycotts of Jewish businesses and their attacks would grow to include disparaging Jewish personalities and communities both at home and abroad and claims of an international Jewish conspiracy against Germany by 1939.

COUGHLIN, THE RADIO PRIEST

Hateful radio content had a genesis that was somewhat earlier and more modest in scope, evolving piecemeal in various places as the power of the new medium became apparent. One early source of hate on the radio that also boasted a global frame of reference can be traced to a small Detroit church known as "the Shrine of the Little Flower" and its priest, Father Charles E. Coughlin.

Coughlin was born in 1891 to an American father and Canadian mother in Hamilton, Ontario. He studied theology at the Catholic St. Michael's College in Toronto and taught for 10 years at Assumption College in Windsor, Ontario. In 1926, Coughlin took charge of a parish in a Detroit suburb and his duties there would eventually lead him to the radio (Krebs, 1979).

Since church services were a common fixture since the earliest days of American broadcast radio, it was no surprise to find Coughlin on the radio. In his case, the nagging financial problems of his church and the tantalizing possibility of being able, through radio sermons, to raise funds to maintain his church (and perhaps pay off its debts) in part motivated his initial foray into radio broadcasting. One account even links his appearance on radio to an incident in which the Ku Klux Klan burned a cross in his churchyard soon after he had assumed his duties there. According to Krebs (1979, p. 44), Coughlin "went to station WJR in Detroit and proposed that he be given air time each Sunday to explain Catholicism to the community."

Biographer Mugglebee (1933, p. 164) described Coughlin's start as taking a leap when he first addressed his "unseen audience on Sunday, October 17, 1926, at two o'clock, Detroit time" and then waited for the response of the sometimes fickle audience. His initial broadcasts starting in 1926 came from a single station, WJR in Detroit; responses were positive (though sparse), and Coughlin continued in a career as a radio preacher. From a total of five letters after his first broadcast, Coughlin's audience response would grow so much that he was reputed to be one of the single largest mail recipients during the 1930s with many of the people on his mailing list of more than two million persons sending him financial contributions. According to Jeansonne (2012, p. 362):

> By the mid-1930s, Coughlin received more mail that anyone in the world. . . . Many sent him small sums that generated a cash flow. Neighborhood children pilfered the priest's trash and found coins and bills overlooked by sorters in their haste.

From a religious start, however, Coughlin's sermons grew into political diatribes and audiences swelled both in support and in opposition to his rhetoric. By 1929 two Chicago stations picked up his program, and within four years of his start, listeners heard Coughlin on 18 stations with the Columbia Broadcasting System with an estimated 40 million tuning in to his program, "The Golden Hour" ("Father Coughlin," 1979, p. 1). Mugglebee (1933, p. 332), a clearly sympathetic writer, while estimating Coughlin's audiences at some 15 million, also acknowledged his detractors and his zealotry:

> In some respects, he supplemented the president; in others, he far overreached him. Of the 15,000,000 listeners who weekly regulated their radio dials to his discourse, there were several millions who disputed

his contentious theories of economics, old as Adam, only clothed to meet modern situations and needs. To them he was a fanatic and zealot and they listened to his "brawling dramatics" to see just how far he could go in his dissertations on "man's inhumanity to man."

Even if Mugglebee's figures are exaggerated (see Lavine and Wechsler's much smaller estimate below), it was true that Coughlin became very influential, in large part due to his radio broadcasts. Coughlin became not just a preacher but also a political commentator, intertwining his religious and political thoughts, blaming bankers and businessmen for society's economic woes and spewing rhetoric that was widely considered anti-Semitic (Mazzenga, 2008; Somerville, 2012). The global dimensions of Coughlin's rhetoric soon became evident as he eventually began voicing support for both Mussolini's Italian Fascism and Hitler's Nazi regime. Lavine and Wechsler (1940, p. 132) considered Coughlin to be Hitler's "most important and most talented" unofficial apologist in the United States, dubbing him the "high priest of American anti-Semitism":

> For many months before the war his weekly sermons and his weekly magazine, *Social Justice,* had been full of thinly veiled apologetics for the Nazis. On several occasions he was apprehended stealing words from Dr. Goebbels' mouth.

Krebs (1979, p. 44) similarly noted that:

> Gradually, Father Coughlin's sermons and his weekly magazine, *Social Justice* which has a circulation of a million, became instruments of anti-Semitism. Units of the Christian Front organization which he supported, made raids on Jewish institutions and businesses. The mere mention of his mane at rallies of the pro-Nazi German-American Bund touched off wild cheering.

After Coughlin called the head of the National Recovery Administration, General Hugh Johnson, "a cream puff general" the general responded with a radio address of his own in which he likened Coughlin to Adolf Hitler, saying:

> You have not chosen the swastika. You have a more sacred device—no swastikas for your Nazi—but a cross. . . . Someone sent me a parallel of what both you and Adolf [Hitler] have proposed and preached and they are as alike as peas in a pod. As a foreign-born you could not be a President but you could be a Reichsfuehrer—just as the Austrian Adolf became a dictator of Germany.

The comparison, though strong, was not completely unwarranted. Coughlin published the anti-Semitic *Protocols of the Elders of Zion* in his magazine and his supporters were known to paint swastikas on Jewish businesses in several cities and to provoke physical altercations with members of the Jewish community in New York (Krebs, 1979). In these ways, Coughlin was clearly reflecting the force of global events of the time, though in a manner that cautions that not all global phenomena are positive ones. Coughlin was kicked off the CBS network in April of 1931 but found his way back on the air later that year on an ad-hoc private network of as many as 47 stations ("Father Coughlin," 1979) cobbled together to facilitate his broadcasts which, in one form or another, continued for several years more.

During his broadcasts Coughlin caused outrage on multiple occasions, such as when he accused two Jewish firms in New York of aiding in the Russian revolution and when he called US President Franklin Roosevelt "the great liar and betrayer" in 1936 ("Father Coughlin," 1979, p. 7). The latter incident raised the ire of his bishop and Coughlin was forced to apologize for his comments. Later in 1936, under what newspapers speculated to be pressure from the Vatican and the US federal government (Krebs, 1979), Coughlin again went off the air as *The Daily News* newspaper (notably, in St. Thomas, US Virgin Islands) reported:

> What Vatican voices have whispered is not known, but the orating ecclesiastic will be heard no longer in politics. His denials of official pressure as a cause make poor evidence. His swan song is: "I will no longer wage the battle I have fought since the general public is indifferent and my fellow clergymen are in flagrant opposition." ("Public Indifferent," 1936, p. 1)

Even after various pressures forced Coughlin off the air, his supporters fought for his return and, of particular interest to the discussion of global radio, pushed for an international hookup so that his message could be more widely dispersed. *Variety* reported in an article datelined Detroit, November 16, 1937:

> If sufficient pressure is brought on church officials by Father Charles E. Coughlin admirers to return radio priest to ozone, it's understood an international hookup will be arranged to carry priest's remarks around the world. . . . Former members of Coughlin's "Social Justice" union have started agitation for priest's return to air and expect to carry appeal to Rome. If the pressure brings fruit, it's reported here Coughlin is ready to shoot the works on an international hookup, blanketing not only the US and Canada, but probably Great Britain and other European countries. ("International Radio: Coughlin," 1937, p. 43)

This international push highlights the fact that, though Coughlin's primary focus was on the United States, his broadcasts were known to have international implications and "his influence reached beyond America and into the realm of international politics" (Cannistraro & Kovaleff, 1971, p. 431). Coughlin was among those who opposed the United States ratifying the Covenant of the League of Nations and joining what was known as the "World Court" and opposed the idea openly in several of his broadcasts. He claimed in early 1935 that "the members of the World Court judiciary are philosophically and nationally prepared to gang us into submission" and warned that an unwanted push to internationalization would visit upon the United States "the debasements of the standardized poverty of Europe" ("Coughlin Opposes," 1935, p. 16). The novelist H. G. Wells (1935, p. 33) described the influence of Coughlin and his audience on the United States' decision against League of Nations and World Court membership despite the best efforts of then President Woodrow Wilson, reporting that the US Senate had been besieged with at least 40,000 telegrams from radio listeners opposing the idea. Father Coughlin and his radio contemporary Will Rogers were primarily responsible for this drive. Wells (1935, pp. 37–38) characterized Coughlin's discourse as xenophobic and anti-Jewish, noting that:

> The empty poverty of his method, considered in relation to his undeniably immense popularity, was a disconcerting symptom to a visitor who still hoped to find in America a practical and moral constructive lead for the rest of the world.

Whether through his radio signals or through news of his exploits, Father Coughlin's fame spread beyond the borders of the United States. When he visited Bermuda in 1936, the *New York Times* correspondent described him as the "famed radio orator" ("Coughlin Is in Bermuda," 1936, p. 16) who planned to both preach at a local church and deliver a political speech during his visit. Coughlin's name was also called in other countries when anti-Semitism reared its head. When a weekly magazine in Manila, Philippines, began publishing anti-Semitic material in 1940, other local publications were forced to speculate on the origins of the material and at least one writer concluded that:

> The fact that the book, *The International Jews* (which originated in the United States), is one of the main sources of "information," rather points to Father Coughlin, "radio-priest" of Detroit. ("Anti-Semitism," 1940, p. 132)

Coughlin would find that his message was eventually not welcome even in the land of his birth, Canada. Though the "Radio Priest" had, at times, enjoyed strong listenership in Canada ("International Radio: Coughlin's Invitation," 1939), Windsor, Ontario's CKLW (one river's width away from Detroit) was among the stations that removed Coughlin's radio programming in 1939 with the suggestion that "there seems no question that Canada has classified Coughlin as pro-Nazi" ("Coughlin Not Okay," 1939, p. 37). Later that same year, government authorities banned Coughlin's magazine from Canada.

While some countries reacted negatively to Coughlin, others took an interest in his activities. We have noted before that NBC carried a New Year's address from Benito Mussolini in 1931. Though this may seem strange to many today, historians Cannistraro and Kovaleff (1971, pp. 427–428) pointed to the somewhat nuanced relationship that the Italian

Fascists enjoyed with the United States prior to US involvement in World War II:

> For many Americans in the 1920s, Mussolini was a popular and pre-stigious figure and Italy a nation worthy of respect. Fascist policymak-ers carefully nurtured pro-Italian sentiment in the United States and consistently sought to appease American public opinion. Mussolini had been counseled to proceed with caution in handling the Americans and consequently the propaganda policies of his government were sub-dued and restrained.

In this context, Coughlin's broadcasts proved to be of great interest to the Italians as did their policies prove attractive to him. According to Cannis-traro and Kovaleff (1971, p. 431), Coughlin geared his programs to match Mussolini's propaganda in hopes of gaining "a useful ally":

> The Radio Priest's activities were carefully followed in Rome since they could serve as a valuable instrument for advertising Italian Fascism in America. At the same time, however, the Italian government recog-nized that his extremist views and radical schemes made open associa-tion with him undesirable.

Thus his multiple attempts to enlist the opinions and endorsement of personalities such as Mussolini (including an invitation to have the Ital-ian leader pen material for Coughlin's magazine) proved fruitless as did his attempts at cooperation with the Nazis in Germany such as "seeking German approval for a document attesting that Nazis supported Chris-tianity" (Jeansonne, 2012, p. 368), which the Germans refused to sign.

A DUBIOUS LEGACY GONE GLOBAL

The tradition of hateful content on the radio in the United States did not end with Coughlin. In varying degrees this thread of discourse has often reared its head both directly and indirectly even on commercial radio and up to relatively recent times. After Coughlin, personalities such as Bob Grant (and his mentor, Joe Pyne, before him) would become famous for provocative styles and aggressive presentations. However, Grant often went beyond the provocative into openly racist and hateful content that garnered both commercial success and notoriety.

Grant was born in 1929 in Chicago and worked in Los Angeles during his early career. He rose to particular fame on New York's WABC, where he started in 1984, by which time he had already been known to feature the Ku Klux Klan's David Duke on his programs (Vitello, 2014). In 1978, when WOR New York fired Grant, *Variety* magazine referred to his spe-cializing "in what many people think of as racial and ethnic slurs in defense of conservative political principles" ("Bob Grant," 1978, p. 34), noting as well that he had referred to a boxing match between two

African-American fighters (Muhammad Ali and Leon Spinks) as "two baboons in a ring" (1978, p. 46). These were not isolated incidents. Norman (1998, p. 92) described several other on-air utterances from Grant:

> Well, on one program . . . Grant said, "We have in our city . . . in our nation, not hundreds of thousands but millions of sub-humanoids, savages, who really would feel more at home careening along . . . the dry deserts of eastern Kenya—people who, for whatever reason, have not become civilized." On another occasion, Grant voiced the following sentiment: "I can't take these screaming savages. Whether they're in that African Methodist Church, A.M.E church, or whether they're in the streets, burning, robbing, looting, I've seen enough of it."

The *Journal of Blacks in Higher Education* called Grant "Talk-Radio's Resident Racist," noting his comments calling for the sterilization of African-Americans and his slurs on Martin Luther King ("Bob Grant," 2014, p. 64).

Under various labels such as "shock," various radio personalities in the United States have furthered the legacies of Coughlin and Grant, gaining national and international prominence through xenophobic positions and statements. Steigler (2014), for example, has identified Michael Savage's (real name Michael Alan Weiner) syndicated talk show *The Savage Nation* (broadcast on some 400 stations) among these and noted that "Michael Savage's proclivity toward hate speech features prominently in the program, and is especially abrasive" (p. 237). This radio personality's international reputation is such that Great Britain banned him from entry under their rule against admitting "hate promoters" (Steigler, 2014). Adams (2009, p. 36) has recounted just one of the incidents that led to that determination:

> "Oh, you're one of the sodomites," he once informed a caller who took issue with his attacks on homosexuals. "You should only get AIDS and die, you pig. How's that? Why don't you see if you can sue me, you pig. You got nothing better than to put me down, you piece of garbage. You have got nothing to do today, go eat a sausage and choke on it."

Noriega and Iribarren (2013) documented the use of derogatory speech toward numerous groups including immigrants and sexual minorities in radio programs including *The Savage Nation* and others such as the *Lou Dobbs Show* and the *Ken & John Show*. While these are all primarily US broadcast features, their content is easily accessible globally on the Internet and form part of the corpus of hateful material that today fuels global hate movements that span national boundaries.

Rwanda

Global radio has been criticized for its roles in propaganda, imperialism, repression and other ills in the past. The negative impacts of radio both local and global continue into the modern age. Legal scholars such as Pauli (2010) have pointed to the role of radio—notably the station *Radio-Télévision Libre des Milles Collines* (*RTLM*) in "orchestrating the Rwandan genocide in 1994" (p. 666). Nor indeed is this the view of one author. The International Criminal Tribunal for Rwanda convicted two *RTLM* executives (along with a local newspaper editor) of inciting genocide through their encouragement of murderous violence.

RTLM was the only privately owned station in the country prior to the genocide, the other station being government owned. For its role in inciting members of the local Hutu population to kill their Tutsi neighbors based on racial hatred, the station was dubbed "Radio Machete." The advantages to which scholars of radio have repeatedly pointed—the accessibility of the medium, the low cost of production and reception, the ability to reach audiences with no need for literacy and the ability to immediately address audiences in their own native languages—were all at play as the Hutu executives used the radio to rouse their kinsmen to slaughter their Tutsi neighbors. According to Mann (1997, p. 5):

> In 1994, the country's main radio station, the *Radio-Television Libre des Milles Collines*, then controlled by Hutu extremists, began broadcasting hate messages targeting members of the rival tribe, the Tutsis, and moderate Hutus. . . . The Rwanda station even broadcast lists of enemies to be hunted down. "Take your spears, clubs, guns, swords, stones, everything, sharpen them, jack them, those enemies, those cockroaches," the station urged listeners. The result was one of the world's worst blood baths, in which more than 500,000 unarmed Tutsis and moderate Hutus were slaughtered.

While this was clearly an internecine conflagration, there are important global dimensions to the role of radio as it played out in Rwanda. One of those dimensions was the fact that the United States was in a position to intervene in these broadcasts. According to Pauli (2010, pp. 666–667):

> As *Radio-Télévision Libre des Milles Collines* (*RTLM*) was orchestrating the Rwandan genocide in 1994, the United States considered jamming its broadcast signals but rejected this form of intervention, citing "international telecommunications law and international conventions regarding the freedoms of information and expression." The virulent broadcasts continued to reach their listeners. In the space of approximately 100 days, in a country of fewer than 8,000,000 people, an estimated 800,000 were killed. RTLM is cited as having incited thousands of civilians with no criminal histories to commit acts of unspeakable brutality.

Pauli (2010) noted, as have others (Cooper, 2001; Metzl, 1997), that the reasons for non-intervention had to do only partly with concern for the international laws and conventions such as the International Telecommunications Union (ITU) rules against jamming. The US decision also had to do with costs, which were estimated at about $8,500 per hour for special airplanes. More to the point, the lack of action on this serious problem had also to do with a lack of international policy and direction from agencies like the United Nations. Indeed, Roméo Dallaire, commander of the United Nations Assistance Mission for Rwanda (UNAMIR), commented on the "crucial role" of both local and international media in the genocide (2007, p. 12):

> I was able to watch the strange dichotomy of local media, on one side, fueling the killing while international media, on the other side, virtually ignored or misunderstood what was happening. The local media, particularly the extremist radio station *Radio-Télévision Libre des Milles Collines (RTLM)*, were literally part of the genocide. The *genocidaires* used the media like a weapon. The haunting image of killers with a machete in one hand and a radio in the other never leaves you. The international media initially affected events by their absence.

Pauli (2010, pp. 669–700) proposed a framework for evaluating the need for foreign intervention into local media attempts at inciting violence that considers, among other factors (such as political instability and pre-existing tensions and violence), the extent to which audiences are insulated from competing messages. In this regard, Straus (2007) has noted that not only were Rwandan audiences able to receive the competing local (government-owned) station known as Radio Rwanda, but they were also able to get news and information from Radio France International, the Voice of America, the British Broadcasting Corporation, and *Deutsche Welle*.

Straus (2007) also argued that claims about the effects of RTLM's broadcasts on the incitement to violence may be somewhat simplistic in that interviews with perpetrators of the violence revealed many other contributing factors beyond the calls on the radio:

> Highlighting modern media is perhaps an easy way to make sense of mind-numbing violence in faraway lands. But to understand how such terrible events occur, we need to look well beyond simplistic frameworks and consider complex issues of agency, context, institutions, and history. Perpetrators of genocide in resource-poor countries are like decision makers elsewhere: they act on the basis of what they see, experience, know, and fear, not simply on the basis of what they hear— or even what they are told—on the radio. (p. 632)

Straus' argument (2007) is consistent with well-established media theory that strongly holds that so-called "media effects" are complex and multi-layered, with messages acting not just directly from mass media sources such as radio to audience members but instead mediated in numerous ways through such filters as the interpretations and responses of personal contacts to these messages or competing narratives from other media and information sources (Klapper, 1960). Writing in 1975, Hale (p. x) identified this lack of competing narratives as a factor both in Hitler's use of radio and its possible future uses toward similar ends:

> Hitler used it deliberately as a way of inducing mass hysteria. There is no reason to suppose that it cannot be used in the same way again wherever in the world the quantity and variety of the local media have not yet induced satiety and skepticism.

However, research has also shown that media can have somewhat more powerful effects when spectacular or extraordinary events are involved that stimulate people to action (Rogers, 2002; Singhal, Rogers & Mahajan, 1999). Among the historical radio landmarks that typified such powerful effects were Orson Welles's 1938 "War of the Worlds" broadcast that sent some audience members scampering to hospitals and police stations for safety and the 1942 Kate Smith war bond drive in which Smith raised over $2,000,000 during a 21-hour radio broadcast.

Returning to the Rwandan situation, these powerful direct effects are also most common when media coverage isolates audiences either because they are the only source of information, or when no alternative source of information on the issue at hand exists. Straus does not address the fact that available foreign or regional radio messages did nothing to counteract the calls to violence. The explanation for this lack of influence and the ability of RTLM to prevail as a primary motivator of action may lie in Dallaire's condemnation of the failure of international media, including the global radio broadcasters who could have reached the Rwandan citizenry to address the situation.

The Rwandan situation demonstrated the use (and misuse) of radio whatever its exact roles and degree of influence. Whereas history has provided numerous examples of international radio and cross-border transmissions being used as harmful or unwanted propaganda, the Rwandan case provides at least an argument for a greater and more purposive role for international radio. The global reach of signals from global radio broadcasters may or may not have been able to influence the situation in Rwanda, but both the established names and emerging global radio voices may be able to exercise greater vigilance and involvement in efforts (or at the very least, voices) to prevent such horrific acts in the future.

Neo-Nazis and Global Currents of Hate

Both globally and locally, the potential harm of radio broadcasting remains a relevant issue, not simply confined to the realm of historical cautions. Boyes (2010), for example, reported on raids in Germany on an Internet radio station calling itself *Widerstand* (Resistance)-Radio and detention of 23 people on charges of spreading racial hatred. The station featured music and other content that included glorification of Nazi atrocities:

> Young neo-Nazi sympathisers have been loading up their MP3 players with poisonous songs and going to work or school tapping their feet to banned lyrics such as "With six million down / that's when the fun begins." The barely veiled reference to murdered Jews is a clear violation of Germany's anti-hate laws but the far Right has gained access to this and other songs from an Internet radio station formally registered in Chicago—a sign that neo-Nazis who glorify the Third Reich are turning to technology to dodge the law. (p. 30)

Perhaps even more chilling in this report was the fact that the hatred being peddled was also international in its production and dissemination, with the United States (with its much more liberal free speech protections) being a source of the material and conceptual propaganda:

> Much of the anti-Semitic pamphlets and tapes produced in the US are smuggled into Germany through a network of mailbox addresses. Music via the Internet, however, is seen as the quickest way into the heads of the young generation. (Boyes, 2010, p. 30)

Using the combined power of the Internet and audio broadcasting for the spread of hate is not a novel strategy, nor is it limited to either Germany or the United States. Such strategies date back to the late 1990s when streaming audio and attempts at Internet radio were still in their relative infancy and could be found even in what might be thought of as unlikely locations such as Canada. Hier (2000) described the so-called "Freedom-Site" that several racial supremacists organizations operated in Toronto. These organizations included groups with names such as "Heritage Front" and the "Euro-Christian Defense League (ECDL)" who hosted, among other content, what Hier (2000, p. 476) described as an "Internet 'radio-station'" that included an archive of white supremacist radio programming as early as 1996. The creator of the site, Marc Lemire, was also leader of the racist hate group known as the Canadian Patriots Network and went on to fight several legal battles in defense of his hate speech activities (Brean, 2011).

Even as the technologies to enable Internet audio streaming and Internet radio broadcasting were still evolving, hate groups were early adopters of the technologies, developing various versions of audio content online. The Anti-Defamation League reported in 1997 that several known hate groups including the Ku Klux Klan and the National Alliance were already adopting and adapting audio and radio to the Internet (Hoffman, 1997, p. 1):

> The KKK has expanded its Web presence as a way of recruiting members to offset its declining influence. David Duke has established a connection with the neo-Nazi National Alliance and has begun to use the Internet extensively and has even started an Internet-only "radio"

program. William Pierce's National Alliance is making extensive use of
Internet radio.

These efforts often leveraged earlier radio broadcasts and sought to surmount restrictions of cost and ratings that stymied efforts to broadcast
hateful messages on traditional broadcast radio. David Duke's efforts to
broadcast on traditional radio failed due to poor ratings, but by providing the same content (some of it recorded from earlier broadcasts) on the
Internet, he boasted of being able to reach a global audience at a fraction
of the cost of traditional radio (Hoffman, 1997).

This globality would be an important element in the development of
hate groups and their use of Internet radio in years to come. Poulter
(2018), for example, has described the connections among a woman
named Dara Leigh Bloom, a neo-Nazi Internet radio station known as the
American Nationalist Network and a British neo-Nazi youth organization calling itself National Action (NA). While supporting the American
Nationalist Network and its radio efforts, Bloom also gave money and
technical support to the British NA, even setting up the group's website
(Poulter, 2018).

Elsewhere in Europe, neo-Nazi organizations such as Norway's Nordic Resistance Movement (NMR) maintain websites that offer online
newspapers and Internet radio services as well as online stores. A hack-
ing attack on NMR's computer systems in 2017 revealed the identities of
Norwegian Nazi sympathizers, but it also allowed investigators in Sweden to expose Swedish users who were using the Norwegian sites to
express Nazi and neo-Nazi sentiments ("Data Leak," 2017). Later, it
would become evident that the cross-border hate activities were not limited to the two Scandinavian neighbors as, in 2018, Norwegian experts
identified connections between the NMR and Russian operatives (a claim
that both the Russian embassy in Oslo and the NMR denied) ("Norway
Experts," 2018; "Russian Embassy," 2018).

In addition to various radio streams, audio content in the form of
podcasts and music are also increasingly popular globally as vehicles of
neo-Nazi propaganda online. Schulberg (2019) described the influence of
these neo-Nazi podcasts among white supremacists (as downloadable
content on streaming outlets with names like Radio Aryan) including the
shooter who killed eleven people at a synagogue in Pittsburgh, Pennsylvania, in 2018. The source of the streaming site Radio Aryan, however, is
not known, with some speculating that it is based in Wales. Radio Aryan
has been known to broadcast readings of Hitler's *Mein Kampf* and defended the gunman who killed people at a mosque in Christchurch, New
Zealand in 2019 (Martin, 2019).

The global connections of neo-Nazi hate do not in any way diminish
the domestic virulence of this scourge that continues today. Well past the
years of Father Coughlin or Bob Grant, reduced costs of operation and

the rise to commercial success of conservative talk radio have more recently enabled the popularity of dedicated hate-filled programming in the United States. Beyond the so-called "shock" programming in its political talk show guise, newer programming such as *The Political Cesspool Radio Show*, which the Southern Poverty Law Center described as "the primary radio nexus of hate in America" (Jenkins A. F., 2008, p. 34), makes Michael Savage and Rush Limbaugh pale by comparison. With few pretensions to mainstream acceptability, shows such as *The Political Cesspool Radio Show* have routinely featured appearances by members of the Ku Klux Klan and neo-Nazis.

In audio form as Internet radio, podcasts and music alongside textual materials, images and videos, modern hatemongers peddle their hateful ideas. Alongside their diatribes and invectives against various others, they sell caps, t-shirts and other merchandise which promote and celebrate their hateful ideologies to local and global audiences. While the technologies are sometimes new, the concepts are not.

Hate in audio content on the air and online also comes in many different forms. Along with the tribal tensions of the Rwanda genocide and the racism of the neo-Nazis, we can also add religious bigotry and hatred. In July 2017, for example, a community radio station in Sheffield in the United Kingdom lost its license for broadcasting more than 25 hours of hateful content from a foreign Islamic preacher and militant whom the United Nations had designated as a terrorist. The station did not deny the broadcast, or that the material was hateful, but rather claimed they did not know the preacher had been designated a terrorist and that they broadcast it because they did not have an announcer on hand ("Radio Preached Hate," 2017). Sometimes the hate on the radio can be a combination of forms, such as the mix of religious and sexual bigotry evident in the 2016 case of an American, Pastor Steven L. Anderson, who used the radio in Botswana to preach that the Bible says that homosexuals were "worthy of death" (Bearak, 2016). Police arrested the pastor at the radio station at the end of his speech and authorities promptly deported him.

At the start of radio, Father Coughlin's speeches had significant international impact. His words and his calls to action influenced policy decisions at home and abroad; his broadcasts were banned in some countries and celebrated in others. A similar set of cross-border dimensions may also be found with modern hate movements. From the cross-border imperatives of the Rwandan genocide to the cross-border ephemerality of the European neo-Nazis and the global aspirations of the Ku Klux Klan's radio operations, the purveyors of hate have consistently sought to spread their message as far as possible and have made radio and audio media key components of their strategies.

NINE
Conclusion

While distance for its own sake was an early driver of radio broadcasting technology, adoption and use, audiences and programmers often placed great pride in their own local content whether meant for their own consumption or beamed (whether directly or indirectly) to distant listeners. This particular global/local dynamic has been evident in the development of local broadcast radio operations within colonial territories and even more so in the postcolonial experience of several nations that have viewed radio as a national resource. Even today, modern commercial radio operations, whether geographically focused or not, recognize the additional markets provided through streaming on the Internet, and some small commercial operators (even in developing countries) are now *de facto* regional and global radio broadcasters through the combination of their terrestrial broadcast operations and networked streaming audio. Such expansions of broadcast reach have also been accompanied, in many cases, by a virtual reduction of distance in which audience members can experience shared listening with others who may be geographically distant and may also be connected with station personalities and other listeners in myriad virtual communities.

EARLY RADIO, MODERN RADIO AND GLOBALIZATION

The evolution of radio as both a technical and a social phenomenon has been an integral part of the global communication processes that we associate with modern human existence through technical changes as vast as the progression from the telegraph to the Internet. Analysis of the global dimensions of its early development reveals that the medium has demonstrated, from its beginnings, an ability to reach global audiences often hurdling barriers of space, language and literacy. In this context,

159

radio has been a globalizing force since well before the advent of the networked digital technologies we emphasize today.

Much greater distances are easily obtained in the age of digital networks with much greater reliability, at lower costs and with the ability to reach interested audiences no matter how widely dispersed. But while newer digital networked technologies have certainly improved on the medium's reach and diversity of offerings, they are by no means the genesis of radio's global reach or influence. As we have seen, amateurs, experimenters and listening pioneers of the early 1920s in far flung corners of the globe reveled in the ability to enjoy transmissions from early broadcasters. Stations in Pittsburgh, Schenectady and Eindhoven were being listed in publications from Georgetown to Capetown as the global scope of the medium became evident, and early broadcasters in Chicago touted their Australian programming relays.

Globalization is a notoriously multifaceted concept viewed through many different prisms in many different fields. Yet, at its core is the notion of bringing groups and individuals together on a global scale whether in terms of integration of financial markets, the exchange of cultural products or the possibilities of communications that surmount distance. The ability of radio, from its very beginnings, to bridge distances and to bring shared experience to geographically dispersed populations was evident in such instances as the use of the medium to support British imperial identity through the BBC and its Empire Broadcasting as well as in the concerns about foreign influences from outside the French Empire among Algerian colonists. Outside of formal ties of empire, hegemonic struggles for global influence were evident, as we have seen, in the propaganda and counter-propaganda efforts of European nations to their neighbors and allies as well as in US and UK attempts to maintain influence in the Caribbean. It should be noted, as well, that such globalization processes were inextricably linked with prior historical globalization processes including European colonial expansion. Notably, Radio Vaticana's global outreach would scarcely have been conceivable if not for the historical spread of Christianity and Catholicism through European conquest and acquisition of lands of which they had been previously unaware.

RADIO, GEOPOLITICS AND DIFFERENT "WORLDS"

The preceding examination of global radio broadcasting in its many shades through time provides some opportunity to examine how the medium and its myriad extensions have been used in different contexts and its various roles in global processes. The picture that emerges is one of a medium with wildly different meanings and implications to different communities of users.

The primary debates surrounding early radio both domestic and global in Europe and the United States were over the question of government regulation and control versus free market enterprise for the medium; there were different concerns in other parts of the world. In colonies of the time such as India, British Guiana and even Algeria, there were very different debates involving whether radio broadcasting was an appropriate or sustainable undertaking for particular colonies given, on the one hand, costs and demands of the medium and, on the other, the pressing need for local radio due to exposure to foreign radio from different (even hostile) global forces. Colonial subjects, however, found the blessing of radio, when acquired, to also be something of a curse. Heavy restrictions ensuring a steady flow of information from imperial authorities on local radio, for example, often prompted radio users to revert to distant signals from Europe or the United States.

In the postcolonial era, the duality of global radio also became a matter of concern as emerging nations felt themselves too easily infiltrated by content from former colonial masters (and other perceived imperialist nations) through global radio. Simultaneously, they felt pressure to respond to such influences by competing with expensive local production of alternative materials and otherwise trying to compete with information flows from outside (such as domestic content quotas instituted in many nations, both developing and developed). Whether with simple economic strategies such as nationalization in the case of Trinidad and Tobago and Guyana or the broader political strategy of arguing against a free flow of information and the right to self-determination of information policy, smaller, developing nations approached the question of global radio with some level of trepidation and at great cost as they attempted to nurse domestic production and transmission operations with public funds. It would thus be economic neoliberal policies favoring not only free trade but also commercial media that would eventually prevail in many developing nations that could ill-afford to prop up inefficient state-owned and -operated radio (and other media) operations. Market forces have even succeeded in cultural development where deliberate government policies have failed, such as the spread of the local dialect in St. Lucia on commercial media and the launch of commercial reggae stations in Jamaica after government liberalized its broadcasting regime.

DISRUPTING ONE-WAY FLOWS

As radio evolved, the means of broadcasting became an important factor in achieving political voice—whether that was to spur imperial solidarity, counter enemy propaganda or, in the case of newly independent nations, establish and protect national identity. Small and less powerful nations sought to insulate themselves against perceived persistent imperialisms and neocolonial influences through strong mass media—usually radio. Even economically developed territories such as Australia and New Zealand have found themselves, at times, disadvantaged on the global stage because of their geographic isolation and needing to fight for a place and a voice in the global interchange of media (Given, 2009). As technology has again changed the nature of radio broadcasting and the Internet has made global access to audiences a reality even for small stations in lesser-developed or geographically isolated countries, classic postcolonial ideas about cultural imperialism and media dominance have tended to somewhat fade.

That is not to say that the problems of inequality or the challenges of development have disappeared, only that they are now analyzed with far greater complexity than the prior one-way flows would have allowed. In a world where radio broadcasts simultaneously form part of a local mediascape and a diasporic experience for migrants in far-off places (Mohammed, 2017; Mohammed & Thombre, 2017), the direction of flow is no longer the key issue; growing more important are the context of the radio messages (and responses to them) and the community of listeners thus created. Thus the concept of participatory media practices has also evolved from the meagre (though often global) feedback of reception reports in the earliest days of radio, to community involvement in projects such as Kothmale in Sri Lanka and audience participation in the Indian *TInka Tinka Sukh* soap opera and the Tanzanian *Twende Na Wakate* series to active involvement in broadcasts (including messaging, song requests and content advice) from local and remote listeners connected to stations via digital networks and social media. While authors such as Algan (2013, p. 87) have primarily focused on the role of modern digital technologies in providing the means of "participatory culture" in radio, one might also argue that these are but the latest manifestations of a continuing process of evolution in which audio broadcasting has sought to emerge as a medium capable of not only reaching the masses but also being reached by them.

The broader context of media studies has also generally strayed away from a focus on the one-way flow of information from developed to developing nations and cast more attention to both burgeoning domestic production in such nations as well as program exchanges and sharing among them. These processes build on the news exchange concepts of the 1970s and 1980s discussed in this text but also capitalize on Internet technologies to distribute programming across stations in multiple territories—often simultaneously. Commercial radio stations in the English-speaking former British colonies of the Caribbean are but one example of developing nations that now routinely interconnect radio programming for broadcast on some of their systems. Radio dials there have, since the 1990s, also featured various stations that have programmed with a regional scope for the Anglophone Caribbean and aimed or relayed signals among the various territories.

LEGACIES AND TRAJECTORIES

Today, the very notion of what constitutes radio is one that is brought into question when streaming music services, for example, claim to be or become conflated with radio (Black, 2001; Coyle, 2000). Regardless of the emergence of new technologies challenging form and function, this is not a new debate. At its every start, broadcast radio did not have a clear definition, nor did it even have a clear name as it was sometimes termed "broadcast audio" to distinguish it from wireless Morse Code signals and point-to-point, two-way voice technology. As we have seen, the term "broadcast" was, also at times, ambiguous and contentious.

What was clear from its very beginnings, however, was the ability of radio to cover vast distances and deliver content almost instantaneously from one point on the globe to another. Whether received directly by powerful antennas or relayed from a dedicated station to local stations, radio was the original medium of mass audiences and mass messages. The fact that it required no special literacies and the availability of increasingly affordable receiving equipment rendered the radio a global medium whether listeners received it through furtive pickup of weak signals from distant continents, through systems of linkages through vast colonial empires or through formal commercial networks of broadcasting stations.

Early radio broadcasting challenged the pre-existing terminologies and sought new terms for emerging concepts. The notion of the radio-phone or radio-telephony or the push for the name radio-casting all suggested both an anchoring to past technologies and a desire to move beyond the constraints of existing terms. Current developments in media technologies urgently call for changes as well. New and emerging global media forms strongly suggest questions over the usage of terms such as "international broadcasting" to describe a particular group of broadcasters engaged in a formalized system of radio with its own peculiarities of genre and style.

Since all broadcasters (or any kind of content producers) who deliver streams of content to the Internet can potentially reach anyone anywhere on the planet with an Internet connection, the designation of international broadcaster now becomes something less clear. Further, since even the earliest broadcasters were quite clearly capable of reaching distant parts of the globe either directly or with the help of repeaters, then they might also be termed "international broadcasters." While it is clear that those involved in the system of international broadcasting represent particular organizations (including government agencies and religious groups) competing in a system of persuasive messaging with international scope, it could reasonably be argued that their designation as international broadcasters now becomes something more akin to a genre label or description of their content and messaging focus.

Such consideration also raises the issue of whether every streaming radio operation can be thought of as international or global simply because it has the theoretical capacity to reach anyone connected to the Internet anywhere. In practical terms, streaming radio is still not necessarily completely global since Internet technology with the bandwidth required for streaming is not universally accessible. The same comment might be made about early radio that required both expertise and equipment to receive distant signals. However, the question of global broadcasting also has to do with the aims of the producers and consumers of content. While broadcasters in Guyana and Trinidad and Tobago deliver their radio streams to the Internet, those streams are received primarily by their diaspora communities in the United States and Canada. Empirical research confirms that most responses to their streams come from persons in the United States and Canada, particularly in urban centers with high concentrations of Caribbean immigrants. These streaming stations, then, are engaged more in an act of broadcast extension than in a global broadcast. Further, none of the radio decision makers interviewed in this and prior research on the streaming operations of radio stations in Trinidad or Guyana indicated any particular programming strategy for their foreign listenership. Indeed, there was unanimity in the notion that their programming maintained a highly local focus even while they streamed their content on the Internet.

The continuing local focus even in the face of global media reach is not a new finding. Early research on Internet use found that web content and audiences were more likely to coalesce around national interest groups rather than to reach outside of them (Halavais, 2000). Similar findings emerged for social media in later years (Mohammed, 2012), and the local-while-distant radio usage of diaspora group fits neatly with this general notion. In the case of ethnic media in the United States, a similar parochial focus may exist. Thus Native American stations, such as the Hopi stations mentioned earlier, use streaming technologies to reach their local audiences and to expand their reach to scattered (diaspora) members of their community.

A part of that local focus is also the language choices of broadcasts and other content. Early radio broadcasts across distant seas and international boundaries in the 1920s became popular, in part, due to their use of music that transcended linguistic barriers. Orchestral presentations of European classical music, for example, were a mainstay of the Eindhoven broadcasts that made the radio listings in the newspapers of the English-speaking British colony of Guiana. Today, language is often a factor in the geographical spread of Internet radio audiences, but one that exerts both positive and negative pressures. While, on the one hand, broadcasts such as the *Diné*-language station from the Navajo Nation in Arizona may draw listeners from that language community who are geographically separated from the Navajo Nation and serve to revitalize and maintain the native language, it also automatically serves to limit the listenership to that small language community—especially for talk and discussion radio. Onuzulike (2014, p. 285) similarly identified the competing dynamics of limited audiences and the contribution of Internet radio sources to the "ethnolinguistic vitality" of the small minority Igbo language from southeastern Nigeria. In this case, radio content on the Internet may reach places in Africa outside of Nigeria where the language is spoken as well as diasporic Igbo speakers in places such as Chicago in the United States. In this way, Internet radio broadcasts and other content may support a language in danger of extinction due to the existence of a declining number of speakers, but also limit potential audience size for the same reason.

The evidence suggests, at least in some cases, that the notion of global reach does not necessarily suggest any particular sense of global appeal or focus, at least in the sense of the traditional international broadcasters with deliberately designed international outreach (often with political or social agendas). The established international broadcasters while switching their operations increasingly away from shortwave to streaming, can make their case for being truly international in a sea of international broadcasting through their continuing focus on news and entertainment of global relevance or interest.

As we have seen before, terms such as "international broadcaster" or even global media are increasingly problematic in a world where all media are global, cross-border phenomena and in which users construct their own preferred reception universe out of numerous options. Their reception universe might include some combination of traditional mass media including broadcast radio and broadcast television (often by satellite or cable) with online media and social media. Through such a constructed reception universe, modern listeners, even in developing countries or rural areas, can engage with their own global media environment. These constructed global mediascapes can be as parochial or as eclectic as the user prefers and thus further complicate the very concept of the international broadcaster.

Even traditional international radio broadcasters themselves are in the process of adapting to the new diversified media options. Not only have several of these traditional international radio broadcasters switched or duplicated many of their signals from shortwave broadcasts to Internet streaming, they have also made efforts to create a local presence in distant places. Thus Radio China International, aware of its potential audiences among the diaspora and ancestral Chinese communities in the United States, proudly announced that:

In addition to our short-wave broadcasts, we also try to make ourselves heard on the local AM and FM frequencies in many parts of the world through different forms of cooperation. We can now be heard locally in cities like Washington, Los Angles, London, as well as dozens of major cities across the world. If you're randomly tuning your AM/FM receiver at your home or in your car, chances are you'll meet us. Just check our Programs section to find out the exact local time and frequencies in your area. ("History and Milestones," 2004)

The BBC similarly responded to what it termed "a dramatic drop in global shortwave listening," estimated to be in the region of some 20 million of their listeners for the 2009/2010 period (BBC Press Office, 2010), by expanding its other platforms including online, mobile and FM services. Rather than focus solely on direct shortwave, which remains an important source of news and entertainment for large numbers of people all over the globe, this veteran service now supplements its traditional shortwave with web services as well as with FM and medium-wave partnerships with local stations to improve its reach and accessibility. The BBC's 2018/2019 plan indicated that the BBC World Service has been in the process of its biggest expansion since the 1940s including transmissions in more than 40 languages and new offices around the world (British Broadcasting Corporation, 2018, pp. 22–23). Their expansion envisages listeners who may be receiving their content "on Digital Radio in the UK, an FM partner in Kenya or the United States or online streaming in Europe" (British Broadcasting Corporation, 2018, p. 23).

CHANGING POLITICS OF RADIO

Radio broadcasting has, over time, served to increase access to information, particularly to those in areas distant from urban centers or those without the literacy skills to utilize print media. Thus, even when radio was a one-way system of mass communication with little chance for feedback or interaction, it demonstrated at least the potential for some democratization of information. The uses and misuses of the medium suggest that national leaders were quick to recognize the radio's potential not just for informing but also for misinforming. Yet even when information control sought to restrict listeners, cross-border and insurgent radio provided alternative information and news of the world. Here too, a democratizing function was evident. In its newer manifestations, coupled with digital global networks providing both unlimited reach to audiences and opportunities for feedback and collaboration, modern globalized radio provides even greater democratic potentials. Though, as we have seen in numerous instances, wherever the potential for democratic uses exists, so does the potential for misuse. While group and participatory listening has been a facet of radio in diverse places for many decades, the decentralizing effect of Internet and mobile networks in combination with radio has enhanced the possibilities of defining groups or communities in much broader geographic terms and to focus on shared values and ideas as the locus of group formation rather than geographical location.

The global scope of radio has made for numerous interesting global exchanges of both positive and negative kinds. From the early cross-empire experiences of listeners in British Guiana tuning in to music and news initially programmed to reach Dutch colonies, to the publication of German broadcast listings including Hitler Youth programming in Indian magazines of the 1930s, global considerations were often inevitable and frequently problematic.

Cautionary Tales

The democratizing and community-building potentials of global radio have never assured positive outcomes. The same power to bridge distances and bring communities closer has always been open to use for quite opposite purposes such as despotism and division. We have examined several examples of the negative potentials of radio here and the ways in which the medium can be used for the spread of prejudice, hate and even worse.

Prejudice and Foreign-Language Broadcasts

While foreign-language broadcasts in the United States served important community-building and distance-bridging functions for immigrant communities with diverse language traditions, these same broadcasts became the subject of suspicion and fear after the United States became involved in World War II. It was precisely the global dimension of such broadcast operations that drew the attention of the government and the Office of War Information, prompting not only the implementation of monitoring and codes of conduct for broadcasters of foreign-language content, but the eventual dissipation of the industry subsector in later years.

The re-emergence of modern foreign-language broadcasting in the United States, particularly commercial Spanish-language radio stations—estimated at over 1,500 today—has often been met with what Casillas (2011, p. 808) has described as "a racialized climate of suspicion" due, in part, to a focus on immigration and minority issues. As a counterpoint to the discourse of globalization and the globalizing potentials of digital networks, therefore, traditional fear, suspicion and "othering" persist in the face of increased global reach and diverse radio offerings. At the very least, such ideas are reminiscent of the complaint of one radio audience member in 1934 who considered it "annoying to tune in on some station and hear a lot of gibberish which one cannot understand coming out of a loudspeaker" (Krysko, 2007, p. 344).

Nazis, Neo-Nazis and Other Hate on Global Radio

Global radio has been blamed for sins of commission and for sins of omission. As we have seen, German broadcasts that reached Asia and South America around World War II were deemed to be dangerous enough to warrant action (even joint action) on the part of the British and US governments to counteract the influence of the German radio stations. In these and other instances, authorities perceived global radio broadcasting as an active threat that required defensive strategies. In other cases, critics have accused global radio broadcasting of failing to act by failing to speak out against harmful and dangerous movements as in the case of Radio Vatican, or by failing to inform the outside world of atrocities and failing to facilitate countermeasures against incitement as in the case of the Rwandan genocide.

The diatribes of Father Coughlin, and more recent instances of neo-Nazi hate speech combined with the power and reach of radio raise the specter of its use as a vehicle of hate. Its power to create community is not intrinsically a positive or a negative power—but one that can be turned to the purposes of its users. Yet, as Pauli (2010) and others have noted, it is within a vacuum of information that hateful broadcasts can have their greatest effects. This argument would appear to press the case for the presence and proliferation of productive, informative and open radio broadcasting both local and global (in its terrestrial, networked, mobile and other forms) to counteract the many available sources of hate and bigotry that have been secondary consequences of increased global information flows.

A FINAL WORD

Scholars have wrestled with radio as a social phenomenon. A grand theory of radio has proven elusive, perhaps with good reason. The medium itself has grown and transformed in often surprising ways. Uncertainty today over the sometimes mercurial forms of digital networked audio and debates over whether such forms as streaming audio services can be legitimately considered radio mirror uncertainties of the past in which wireless, well established as a form of telegraphy, emerged into contested forms and terms such as "radio telephony," whose forms and functions were often a matter of social and technical negotiation.

Making the matter more complicated has been the global/local duality of the medium which has played diverse roles with audiences at home and in distant places. The spread of global networked technologies has not only expanded the reach of radio stations and services, but also created a renewed focus on the community of listeners who may be geographically dispersed individuals in faraway places or who may be organized locally into listener groups. In any variation of listener circumstances, combinations of radio and networked digital technologies provide at least the potential for virtual communities of interest.

It is perhaps in this concept of community that global radio, from its beginnings as distant stray signals and shortwave broadcasts to its current manifestations of streaming (with social media links), provides the most fertile ground for analysis. In each act of listening, audience members have, to lesser or greater extents, engaged in a community activity. This is clear when these listeners have been villagers around a communal set in rural India but probably equally true when the listener has been a solitary figure in the jungles of British Guiana joining in with an audience for classical music in the United States or Great Britain. When ancestral Indians in Trinidad and Tobago tuned in to All India Radio to listen to broadcasts from India, they also engaged in an imagined community of listeners.

Radio created communities of listeners among early enthusiasts that spread across great distances. Those communities took pride in being able to receive the most distant signals and shared their listening achievements with others. In 1933, for example, E. S. Christiani Jr. of Georgetown, British Guiana, proudly shared a list of best bets for station reception using a Philco Model 43H receiver including BBC frequencies, American stations, German stations and one from Morocco ("Reception in British Guiana," 1933). This example also reminds us that global radio was used to maintain community within empires including the British and Dutch colonial systems, whose programmers often deliberately shored up imperial sentiment and spirit through radio broadcasting.

Global radio, from its beginning to the development of shortwave to the institution of streaming, has thus enabled the formation of communities and participation in such communities for producers and listeners all around the world. In this way, it has been a persistent tool of globalization and provides one of the most promising avenues (with all its limitations and threats) for global understanding.

References

A brief history of broadcasting in Guyana. (2010, October 1). *The Guyana Review*, p. 14.

A notable broadcast. (1928, August 1). *Modern Wireless*, pp. 155–156.

About us. (2018, December 31). Retrieved from Ogun Radio: https://ogunradio.ng/aboutus/.

About us: KAZN AM1300 Chinese Radio Station. (2019, April 30). Retrieved from KAZN AM1300 Chinese Radio Station: http://www.am1300.com.

Abraham, R. (2012). India and its diaspora in the Arab Gulf countries: Tapping into effective "soft power" and related public diplomacy. *Diaspora Studies, 5*(2), 124–146.

Adams, G. (2009, May 9). Mr. Angry; The Saturday profile: Michael Savage. *The Independent*, p. 36.

Adams, M. (2012). *Lee de Forest: King of radio, television, and film.* New York, NY: Copernicus Books/Springer Science+Business Media. doi:10.1007/978-1-4614-0418-7.

Adelson, A. (1994, May 9). Radio stations reduce coverage of local news. *New York Times*, p. D8.

Adler, J. (2004). The "sin of omission"? Radio Vatican and the anti-Nazi struggle, 1940–1942. *Australian Journal of Politics & History, 50*(3), 396–406. doi:10.1111/j.1467-8497.2004.00342.x.

A.I.R. programs overseas. (1938, November 29). *The Times of India*, p. 19.

Aïtel, F. (2013). Between Algeria and France: The origins of the Berber movement. *French Cultural Studies, 24*(1), 63–76. doi:10.1177/0957155812464150.

Aitken, H. G. (1985). *Syntony and spark: The origins of radio.* Princeton, NJ: Princeton University Press.

Akbar, M. (1946, January 11). Letter to Kamaluddin Mohammed. *Kamaluddin Mohammed Papers. Correspondence. January 1946–October 1948, Box 1(Folder 5).* St. Augustine, Trinidad and Tobago: Special Collections, West Indian, The Alma Jordan Library.

Akbar, M. (1947, August 30). Letter to Kamaluddin Mohammed. *Correspondence. January 1946–October 1948, Box 1(Folder 5).* St. Augustine, Trinidad and Tobago: Special Collections, West Indian, The Alma Jordan Library.

Alderton, D., Krim, K., Schmitt, J., & Sheehy, F. (1999, March 1). Digital what? The coming revolution in radio. *McKinsey Quarterly*, pp. 124–129.

Alexander, G. (2008, August 1). The future of black radio. *Black Enterprise, 39*(1), 104–111.

Algan, E. (2013). Youth, new media and radio: Mobile phone and local radio convergence in Turkey. In J. Loviglio & M. Hilmes (Eds.), *Radio's new wave: Global sound in the digital era* (pp. 79–90). New York: Routledge.

Algeria hears Chelmsford. (1924, September 2). *The Christian Science Monitor*, p. 6.

Algiers Correspondent. (1931, March 27). Dominion and foreign broadcasting intelligence: North Africa—weekly English talks. *World Radio*, p. 490.

Algül, F. (2013). An Internet radio from Turkey as an example of community radio: Nor Radyo. *AJIT-e: Online Academic Journal of Information Technology, 4*(12), 75–103. doi:10.5824/1309-1581.2013.2.004.x.

All India Radio. (1940). *Report on the progress of broadcasting in India.* Delhi: Manager of Publications.

Amateur broadcast station opened. (1935, February 5). *The Daily Argosy*, p. 5.

Amateur radio in new laurels of achievement: S. Africa and U. S. in 2 way communication. (1925, November 8). *Chicago Daily Tribune*, p. C8.

Ambrose, R. (1998, May 1). Radio Zulu. *The Beat, 17*(3), 32–36.

American radio to span the globe: Twenty million dollar company formed, with General Electric in it. (1920, January 5). *New York Times*, p. 1.

Anderson, B. (1983). *Imagined communities: Reflections on the origin and spread of nationalism*. London: Verso.

Anti-Semitism in the Philippines. (1940, April 1). *Philippine Magazine, 37*(4), 132.

Arceneaux, N. (2012). In search of alien aerials: The World War I campaign against amateur radio. *Journalism History, 38*(1), 2–12.

Arnheim, R., & Collins-Bayne, M. (1941). Foreign language broadcast over local American stations. In P. F. Lazarsfeld & F. Stanton (Eds.), *Radio Research* (pp. 3–57). New York: Duell, Sloan & Pearce.

At deadline: Ethnic news network set. (1969, March 10). *Broadcasting, 76*(10), 9.

Atkinson, C. F. (1931, March 21). The International Broadcasting Union. *The Spectator*, p. 8.

Austin, C. (1922, May 1). The romance of the radio telephone. *Radio Broadcast, 1*(1), 9–20.

Axis contradictions over India. (1942, April 9). *The Times*, p. 4.

Bagdikian, B. H. (2004). *The new media monopoly*. Boston, MA: Beacon.

Banerjea, S. (1922, October 1). Radio communication in India. *The Wireless Age, 20*(1), 53–54.

Banerjea, S. (1923, August 1). India studies radio broadcasting. *The Wireless Age*, p. 77.

Baraka, R. (2001, October 13). American Urban Radio Networks: For three decades, the nation's foremost black radio network has informed, entertained and served its community. *Billboard—The International Newsweekly of Music, Video and Home Entertainment, 113*(41), 25, 32.

Barnouw, E. (1966). *A tower in Babel: A history of broadcasting in the United States* (Vol. I). New York: Oxford University Press.

Barnouw, E. (1968). *The golden web: A history of broadcasting in the United States* (Vol. II). New York: Oxford University Press.

Baruah, U. L. (1983). *This is All India Radio: A handbook of radio broadcasting in India*. New Delhi: Publications Division, Ministry of Information and Broadcasting, Govt. of India.

BBC Press Office. (2010, July 5). BBC World Service attracting millions in new audiences to TV, online, mobile and FM in changing media climate, says 2009/10 annual review. *BBC Press Office*. London, United Kingdom: British Broadcasting Corporation. Retrieved from http://www.bbc.co.uk/pressoffice/pressreleases/stories/2010/07_july/05/ws_review.shtml.

Bearak, M. (2016, September 20). Botswana deports American anti-gay pastor after radio hate speech. *The Washington Post*. Retrieved from https://www.washingtonpost.com/news/worldviews/wp/2016/09/20/botswana-deports-american-anti-gay-pastor-after-radio-hate-speech/?utm_term=.e98714dbff33.

Beckford, G. (1972). *Persistent poverty: Underdevelopment in plantation economies of the Third World*. Oxford: Oxford University Press.

Bent, S. (1937). International broadcasting. *The Public Opinion Quarterly, 1*(3), 117–121.

Berg, J. S. (2013). *The early shortwave stations: A broadcasting history through 1945*. Jefferson, NC & London: McFarland & Company Inc.

Better wireless programmes. (1927, January 02). *The Sunday Times*, p. 14.

Bhatnagar, S., Dewan, A., Torres, M. M., & Kanungo, P. (2003). *Sri Lanka's Kothmale community radio Internet project (English): Empowerment case studies*. World Bank. Washington, DC: World Bank. Retrieved from http://documents.worldbank.org/curated/en/952191468115460578/Sri-Lankas-Kothmale-Community-Radio-Internet-Project.

Birdsall, C. (2012). *Nazi soundscapes: Sound, technology and urban space in Germany, 1933–1945*. Amsterdam, Netherlands: Amsterdam University Press.

Black, D. A. (2001). Internet radio: A case study in medium specificity. *Media, Culture & Society, 23*(3), 397–408. doi:10.1177/016344301023003007.

Blackwell, G. (1998, June 18). MP3 taking Net world by storm. *The Toronto Star*, p. K1.

Bob Grant, a pain in the latenight, yanked by WOR. (1978, November 8). *Variety*, pp. 34, 46.

Bob Grant: Talk-radio's resident racist. (2014). *The Journal of Blacks in Higher Education, 44*(1), 64. doi:10.2307/4133744.

Bondyopadhyay, P. K. (1998). Sir J. C. Bose's diode detector received Marconi's first transatlantic wireless signal of December 1901 (The "Italian navy coherer" scandal revisited). *Proceedings of the IEEE, 86*(1), 259–285. doi:10.1109/5.65877.

Bonini, T. (2011). The media as "home-making" tools: Life story of a Filipino migrant in Milan. *Media, Culture & Society, 33*(6), 869–883. doi:10.1177/0163443711411006.

Boord, K. R. (1944, December 1). International short-wave. *Radio News*, pp. 54, 132.

Boord, K. R. (1945, January 1). International short-wave. *Radio News*, pp. 58–59.

Boord, K. R. (1947, July 1). International short wave. *Radio News*, pp. 64–66.

Boord, K. R. (1947, December 1). International short-wave. *Radio News*, pp. 65, 149.

Borzillo, C. (1994, April 16). Networks and syndication: Pinpointing users of syndication. *Billboard*, pp. 86, 88.

Boyd, D. A. (1999a). *Broadcasting in the Arab world: A survey of the electronic media in the Middle East* (3rd ed.). Ames, IA: Iowa State University Press.

Boyd, D. A. (1999b). Hebrew-language clandestine radio broadcasting during the British Palestine mandate. *Journal of Radio Studies, 6*(1), 101–115. doi:10.1080/1937652 9909391711.

Boyes, R. (2010, November 5). Raids on neo-Nazi radio silence the airwaves of hate. *Times of London*, p. 30.

Bradley, I. (1981, February 27). Overseas: Soviet jammers face increase in radio broadcasts by BBC. *The Times*, p. 7.

Bratu, R. (2014). Portrayals of Romanian migrants in ethnic media from Italy. *Journal of Comparative Research in Anthropology and Sociology, 5*(2), 199–217.

Brean, J. (2011, December 10). Hate speech battle headed for showdown. *National Post (Canada)*, p. A1.

Brereton, B. (2007). Contesting the past: Narratives of Trinidad & Tobago history. *NWIG: New West Indian Guide, 81*(3/4), 169–196.

Briggs, A. (1961). *The history of broadcasting in the United Kingdom: The birth of broadcasting* (Vol. I). London: Oxford University Press.

Briggs, A. (1995). *The history of broadcasting in the United Kingdom: The golden age of wireless* (Vol. II). Oxford: Oxford University Press.

Bringing India to you. (2000, March 31). *India Currents, 13*(12), 24.

Britain's Broadcasting Service for its empire. (1935, November 1). *Wireless Magazine*, p. 291.

British again fail in around world broadcast effort. (1924, April 27). *Chicago Daily Tribune*, p. J14.

British Broadcasting Corporation. (1926, July 15). *The Times*, p. 27.

British Broadcasting Corporation. (2018). *BBC annual plan 2018/19*. London: British Broadcasting Corporation. Retrieved from http://downloads.bbc.co.uk/aboutthe bbc/insidethebbc/howwework/reports/pdf/bbc_annual_plan_2018.pdf.

British strike proves worth of broadcasting. (1926, June 13). *Chicago Sunday Tribune*, p. 8.

Brittain, J. E. (1992). *Alexanderson: Pioneer in American electrical engineering*. Baltimore: Johns Hopkins University Press.

Broadcast radio telephoning: Westinghouse covering all the country. (1922, January 1). *Hardware Dealers' Magazine*, p. 136.

Broadcasting. (1929, December 14). *The Chicago Defender*, p. 7.

Broadcasting in Germany. (1927, October 6). *The Times*, p. 6.

Broadcasting in India. (1940, November 30). *Nature*, pp. 709–711. doi:10.1038/146709a0.

Broadcasting in India: Not long now. (1923, February 26). *The Times of India*, p. 13.

Broadcasting: Early start possible. (1922, September 1). *The Cologne Post: A Daily Newspaper Published by the Army of the Rhine*, p. 4.

Broadcasting: The community set. (1937, March 23). *The Times*, p. 51.

Broadcasting: The Delhi conference. (1923, March 8). *The Times of India*, p. 10.

Bronfman, A. (2013). El octupus acustico: Broadcasting and empire in the Caribbean. In J. Loviglio & M. Hilmes (Eds.), *Radio's new wave: Global sound in the digital era* (pp. 147–162). New York, NY: Routledge.

Bronfman, A. (2016). *Isles of noise: Sonic media in the Caribbean*. Chapel Hill, NC: University of North Carolina Press.

Brown, A. (1995). Towards regionalization of new communication services in the CARICOM: A technological free-for-all. *Canadian Journal of Communication, 20*(3), 301–315.

Brown, M. (2005). Radio Mars: The transformation of Marconi's popular image, 1919–1922. In J. E. Winn & S. L. Brinson (Eds.), *Transmitting the past: Historical and cultural perspectives on broadcasting* (pp. 16–33). Tuscaloosa, AL: University of Alabama Press.

Browne, D. R. (1982). *International radio broadcasting: The limits of the limitless medium*. New York, NY: Praeger.

Bush wants to boost Middle East broadcasts. (1990, September 17). *Broadcasting*, p. 69.

Business men of Broadway hear race lecturer: Dr. Hubert Harrison talks to big business in financial district on "The Negro and the nation." (1923, July 7). *New Journal and Guide*, pp. 1, 6.

Cairncross, F. (1997). *The death of distance: How the communications revolution will change our lives*. Boston, MA: Harvard Business School Press.

Cambridge, V. C. (2015). *Musical life in Guyana: History and politics of controlling creativity*. Jackson, MS: University Press of Mississippi.

Cannistraro, P. V., & Kovaleff, T. P. (1971). Father Coughlin and Mussolini: Impossible allies. *Journal of Church and State, 13*(3), 427–443.

Caribbean Family Planning Affiliation. (1995). *Annual report*. St. John's, Antigua: Caribbean Family Planning Affiliation.

Casillas, D. I. (2011). Sounds of surveillance: U.S. Spanish-language radio patrols La Migra. *American Quarterly, 63*(3), 807–829.

CBS Radio gets back in shortwave. (1968, November 4). *Broadcasting*, p. 77.

Celler would curb radio advertising. (1924, March 24). *New York Times*, p. 24.

Central Intelligence Agency. (2007, May 29). *A look back. . . . The National Committee for Free Europe, 1949*. Retrieved from Central Intelligence Agency: https://www.cia.gov/news-information/featured-story-archive/2007-featured-story-archive/a-look-back.html.

Chandisingh, R. (1983). The state, the economy, and type of rule in Guyana: An assessment of Guyana's "socialist revolution." *Latin American Perspectives, 10*(4), 59–74. doi:10.1177/0094582X8301000405.

Charlesworth, H. (1935, January 1). Broadcasting in Canada. *The Annals of the American Academy of Political and Social Sciences, 177*(1), 42–48. doi:10.1177/000271623517700106.

Chevaldonné, F. (1988). Nationalization, market economy and sociocultural development: The structures of audiovisual communication in independent Algeria. *Media Culture and Society, 10*(3), 269–284. doi:10.1177/016344388010003002.

Chicago station covers 9,670 miles to Tasmania. (1924, April 27). *New York Times*, p. 17xx.

Chin, G. (2009, July 26). The rise and fall of Guyana's cinemas. *Starbroek News*, p. 1. Retrieved from http://www.stabroeknews.com/2009/archives/07/26/the-rise-and-fall-of-guyana%E2%80%99s-cinemas/.

Chiumbu, S. (2009). *Public broadcasting in Africa: Zimbabwe*. Johannesburg, South Africa: Open Society Foundations.

Chua, A. L. (2012). "The modern magic carpet": Wireless radio in interwar colonial Singapore. *Modern Asian Studies, 46*(1), 167–191. doi:10.1017/S0026749X11000618.

Churches to use radio to reach millions. (1923, December 27). *New York Times*, p. 5.

Cianfarra, C. M. (1944). *The Vatican and the war.* New York, NY: American Book-Straford Press Inc.

Closed circuit: Louder and longer. (1983, April 11). *Broadcasting, 104*(15), 7.

Cockaigne. (1924, November 1). A London letter. *Billboard, 36*(44), 37.

Cohen, L. (1990). *Making a new deal: Industrial workers in Chicago, 1919–1939.* New York, NY: Cambridge University Press.

Comedy troupe tours city. (1927, February 4). *New Daily Chronicle,* p. 4.

Commonwealth Broadcasting Association. (1988). *Handbook—Commonwealth Broadcasting Association.* London: Secretariat of the Commonwealth Broadcasting Association.

Cooper, G. (2001, August 23). Memos reveal Rwanda delay. *Washington Post,* p. A20.

Coughlin is in Bermuda. (1936, November 24). *New York Times,* p. 16.

Coughlin not okay with Canada. (1939, September 20). *Variety,* p. 37.

Coughlin opposes World Court entry. (1935, January 21). *New York Times,* p. 16.

Cox, J. (2009). *American radio networks: A history.* Jefferson, NC: McFarland & Company Inc.

Coyle, R. (2000). Digitising the wireless: Observations from an experiment in "Internet Radio." *Convergence: The International Journal of Research into New Media Technologies, 6*(3), 57–75. doi:10.1177/135485650000600305.

Crisell, A. (1997). *An introductory history of British broadcasting.* London: Routledge.

Current topics: The radio in Egypt. (1933). *The Muslim World/The Moslem World, 23*(2), 195–196. doi:10.1111/j.1478-1913.1933.tb00252.x.

Cuthbert, M. (1990). A model for a small independent news agancy: CANA and CANA radio. In S. H. Surlin & W. C. Soderland (Eds.), *Mass media and the Caribbean* (pp. 403–413). New York, NY: Gordon and Breach.

Czech protest: RFE is "warmongering." (1951, June 4). *Broadcasting, Telecasting, 40*(23), 55.

Daily Chronicle. (1948). *Who's who in British Guiana: 1945–1948* (4th ed.). Georgetown, Guyana: Daily Chronicle Ltd.

Dallaire, R. (2007). The media dichotomy. In A. Thompson (Ed.), *The media and the Rwanda genocide* (pp. 12–19). Ottawa, Canada: International Development Research Centre.

Damome, E. L. (2011). The community of radio listeners in the era of the Internet in Africa: New forms and new radio content, the fan club Zephyr Lome (Togo) as a basis for analysis. In A. Gazi, G. Starkey & S. Jedrzejewski (Eds.), *Radio content in the digital age: The evolution of a sound medium* (pp. 237–246). Chicago, IL: Intellect, The University of Chicago Press.

Darrah, D. (1929, June 9). Americans help Pope build new Vatican state: Furnish cash for phone and radio systems. *Chicago Daily Tribune,* p. 18.

Data leak exposes details of Norwegian neo-Nazi group. (2017, September 22). *BBC Monitoring Europe,* p. 1.

Deutsch, K. W. (1953). *Nationalism and social communication: An inquiry into the foundations of nationality.* Cambridge, MA: MIT Press.

Deutsche Welle. (1982). *DW Handbuch für internationalen Kurzwellenrundfunk.* Berlin: Verlag Spiess.

Distribution: BBC release compact disc. (1984, August 24). *Broadcast,* p. 76.

Domatob, J. K. (1985). Radio Cameroun and rural exodus: Policies and problems. *Gazette, 36*(2), 121–137. doi:10.1177/001654928503600204.

Domatob, J. K., & Hall, S. W. (1983). (1983). Development journalism in black Africa. *Gazette, 31*(1), 9–33. doi:10.1177/001654928303100102.

Dominion and foreign broadcasting intelligence: Foreign music from Radio-Toulouse. (1928, February 24). *World-Radio,* p. 269.

Drama by radio: Broadcasting in America. (1923, March 09). *The Times of India,* p. 14.

Dukepoo, C. (2013). Wishing on "shooting stars" Hopi radio reignites a culture and its language. *Cultural Survival Quarterly, 37*(1), 22–23.

Dunn, H. (2014). Imperial foundations of 20th-century media systems in the Caribbean. *Critical Arts, 28*(6), 938–957. doi:10.1080/02560046.2014.990644.

Echoes and memories: An early experimenter looks back. (1937, February 2). *The Times of India*, p. 16.

Education by radio in India. (1932, June 29). *The Times*, p. 10.

Egypt censors radio broadcasts to outside. (1957, April 14). *Chicago Daily Tribune*, p. 30.

Egypt comes of age. (1938, February 1). *The Spur, 61*(2), 25, 48.

Egypt heard from. (1935, April 3). *Variety, 3*, 41.

Egyptian radio propaganda. (1955, November 8). *The Times*, p. 7.

Eldin, Z. (1941, June 1). Inside Egypt. *Current History and Forum, 53*(1), 32–34, 46–47.

Electioneering by radio. (1921, October 29). *Electrical Review, 79*(18), 673.

Empire broadcasting. (1933, January 17). *Nature*, pp. 16–17.

Equipment & engineering: The many-faceted legacy of radio's Ernst Alexanderson. (1975, May 26). *Broadcasting, 88*(21), 44.

Estill, A. K. (1946, January 10). U.S.-British radio: Two nations broadcast newspaper every day to Caribbean neighbors. *The Wall Street Journal*, p. 1.

Ethnic group programming expands. (1962, September 3). *Broadcasting, 63*(11), 42.

Fahie, J. J. (1901). *A history of wireless telegraphy: Including some bare-wire proposals for subaqueous telegraphs* (2nd ed.). Edinburgh: William Blackwood and Sons.

Fandy, M. (2007). *(Un)civil war of words: Media and politics in the Arab world.* Westport, CT: Praeger Publishers.

Fanon, F. (1961, 2004). *The wretched of the earth.* (C. Farrington & R. Philcox, Trans.) New York: Grove Press.

Fanon, F. (1970). *A dying colonialism.* Hammondsworth: Pelican.

Fargus, B. C. (1953, January 14). Letter from Trinidad Broadcasting Company to Norman L. Maguire. Port-of-Spain, Trinidad & Tobago. Retrieved from https://www.americanradiohistory.com/hd2/IDX-Short-Wave/Veries-IDX/IDX/Foreign-OCR-Page-0136.pdf.

Father Coughlin, "radio priest" for millions in 30's, dies at 88. (1979, October 28). *The Michigan Daily*, pp. 1, 7.

Fawcett, A. (1927, January 8). Band programme. *New Daily Chronicle*, p. 5.

Fay, J. (1999). Casualties of war: The decline of foreign language broadcasting during WWII. *Journal of Radio Studies, 6*(1), 62–80.

Federal Communications Commission. (1941). *Report on chain broadcasting.* Washington, DC: Federal Communications Commission.

Federal Communications Commission. (2003–2004). *A short history of radio.* Retrieved from Federal Communications Commission: https://transition.fcc.gov/omd/history/radio/documents/short_history.pdf.

Fejes, F. (1986). *Imperialism, media, and the good neighbor: New deal foreign policy and the United States shortwave broadcasting to Latin America.* Norwood, NJ: Ablex Publishing Corporation.

Ferretti, F. (1970). The white captivity of black radio. *Columbia Journalism Review, 9*(2), 35–39.

Fessenden, H. M. (1940). *Fessenden: Builder of tomorrows.* New York, NY: Coward-McCann, Inc.

Fiedler, A., & Frère, M.-S. (2016). "Radio France Internationale" and "Deutsche Welle" in Francophone Africa: International broadcasters in a time of change. *Communication, Culture & Critique, 9*(1), 68–85. doi:10.1111/cccr.1213.

Folami, A. N. (2010). Deliberative democracy on the air: Reinvigorate localism—resuscitate radio's subversive past. *Federal Communications Law Journal, 63*(1), 141–194.

Food and Agricultural Organization of the United Nations. (1997). *Nutrition education for the public.* Rome: FAO.

Forde, S., Meadows, M., & Foxwell, K. (2009). *Developing dialogues: Indigenous and ethnic community broadcasting in Australia.* Bristol, UK: Intellect Books Ltd.

Foreign news. (1936, March 1). *All-Wave Radio*, pp. 128, 144.

Fortner, R. S. (2005). *Radio, morality, and culture: Britain, Canada and the United States.* Carbondale: Southern Illinois University Press.

France bans advertising on all radio programs. (1934, December 29). *The China Press*, p. 1.

Free and unfettered. (1979, January 12). *Guyana Chronicle*, p. 6.

Frequently asked questions. (2018, January 20). Retrieved from Radio Africa Online: http://soukous.org/faq.htm.

G. A. Natesan & Co. (1921). *Sir Jagadish Chander Bose, his life, discoveries and writings.* Madras, India.

Garber, K. (1990, April 24). Westwood One announces first commercial sponser of its Soviet radio programming. *Business Wire*, p. 1.

Garratt, G. (1994). *The early history of radio: From Faraday to Marconi.* Stevenage: The Institution of Engineering and Technology. doi:10.1049/PBHT020E.

Gaskins, A. (2009). Let U.S. prey: Mark Twain and Hubert Harrison on religion and empire. *The Journal of Transnational American Studies, 1*(1), 57–61.

Germans in Africa. (1967, October 7). *The Chicago Defender*, p. 11.

Ghany, H. (1996). *Kamal: A lifetime of politics, religion and culture.* San Juan: Kamaluddin Mohammed.

Given, J. (2009). Another kind of empire: The Voice of Australia, 1931–1939. *Historical Journal of Film, Radio & Television, 29*(1), 41–56. doi:10.1080/01439680902722584.

Goebbels, J. (1933, August 18). *The radio as the eighth great power.* Retrieved from Calvin College German Propaganda Archive: http://www.calvin.edu/academic/cas/gpa/goeb56.htm.

Goodman, D. (2016). A transnational history of radio listening groups II: Canada, Australia and the world. *Historical Journal of Film, Radio and Television, 36*(4), 627–648. doi:10.1080/01439685.2015.1134118.

Goodman, D., & Smulyan, S. (2013). Portia faces the world: Re-writing and re-voicing American radio for an international market. In J. Loviglio & M. Hilmes (Eds.), *Radio's new wave: Global sound in the digital era* (pp. 163–179). New York, NY: Routledge.

Gordon, N. (2008). *Media and the politics of culture: The case of television privatization and media globalization in Jamaica (1990–2007).* Boca Raton, FL: Universal Publishers.

Government at Peking plans broadcasting. (1925, February 12). *The China Press*, p. 1.

Gower, F. (1939, February 24). Broadcasting in Germany. *The Spectator*, pp. 294–295.

Grandguillaume, G. (1998). Arabisation et légitimité politique en Algérie. In S. Chaker (Ed.), *Langues et pouvoir: de l'Afrique du Nord à l'Extrême-Orient* (pp. 17–23). Paris: Edisud.

Granville, W. K. (1981, December 23). Action line. *Chicago Tribune*, p. A10.

Graves, H. N. (1941). *War on the short wave.* New York, NY: Foreign Policy Association/ Headline Books.

Greswell, W. H. (1893). *Outlines of British colonisation.* London, UK: Percival and Co.

Gumucio Dagron, A. (2001). *Making waves: Stories of participatory communication for social change.* New York: The Rockefeller Foundation.

Haiti's chief asks U.S. bar radio program. (1956, September 2). *Chicago Tribune*, p. D10.

Halavais, A. (2000). National borders on the World Wide Web. *New Media and Society, 2*(1), 7–28. doi:10.1177/14614440022225689.

Hale, J. (1975). *Radio power: Propaganda and international broadcasting.* Philadelphia: Temple University Press.

Hall, S. (1989). Cultural identity and cinematic representation. *Framework, 36*(1), 68–81.

Hall, S. (1997, March 12). *Representation and the media.* Retrieved from Media Education Foundation: http://www.mediaed.org/assets/products/409/.

Hall, S. (1997). *Representation: Cultural representations and signifying practices.* (S. Hall, Ed.) London: Sage.

Hanoomansingh, H. (2015, July 15). Interview with author. (S. N. Mohammed, interviewer).

Hauser, M. W., Delle, J. A., & Armstrong, D. V. (2011). Historical archaeology in Jamaica. In M. W. Hauser, J. A. Delle & D. V. Armstrong (Eds.), *Out of many, one people: The historical archaeology of colonial Jamaica* (pp. 1–22). Tuscaloosa: The University of Alabama Press.

Hayes, J. E. (2018). Community media and translocalism in Latin America: Cultural production at a Mexican community radio station. *Media, Culture & Society, 40*(2), 267–284. doi:10.1177/0163443717693682.

Hayes, R. (2011). *Subhas Chandra Bose in Nazi Germany: Politics, intelligence and propaganda 1941–43.* Oxford, UK: Oxford University Press.

Hears KDKA on one tube in British Guiana jungle. (1924, November 9). *Chicago Daily Tribune*, p. F 10.

Heid, J. (2000, April 7). How it works: Streaming audio. *PC World.Com*, p. 1.

Heingartner, D. (2007, March 5). Patent fights are a legacy of MP3's tangled origins. *The New York Times*, p. C3.

Henderson, R. (2015, July 15). The origins of 103FM. (S. N. Mohammed, interviewer).

Hendy, D. (2000). *Radio in the global age.* Cambridge, UK: Polity Press.

Here and there. (1924, May 18). *Chicago Daily Tribune*, p. F12.

Hey Dad, remember Trinidad? (1985, September 1). *Popular Communications*, p. 19.

Hier, S. P. (2000). The contemporary structure of Canadian racial supremacism: Networks, strategies and new technologies. *The Canadian Journal of Sociology, 25*(4), 471–494. doi:10.2307/3341609.

Hillenkoetter, R. H. (1948, August 26). *US government officials discuss émigré broadcasts to eastern Europe.* Retrieved from The Wilson Center: http://digitalarchive.wilsoncenter.org/document/114321.

Hiller, H. H., & Franz, T. M. (2004). New ties, old ties and lost ties: The use of the internet in diaspora. *New Media and Society, 6*(6), 731–752. doi:10.1177/14614480 4044327.

Hilliard, R. L., & Keith, M. C. (2005). *The quieted voice: The rise and demise of localism in American radio.* Carbondale, IL: Southern Illinois University Press.

Hilmes, M. (1997). *Radio voices: American broadcasting, 1922–1952.* Minneapolis, MN: University of Minnesota Press.

Hinds, J. (1936, March 1). Globe girdling. *All-Wave Radio, 2*(3), 112–115.

History and milestones. (2004, December 31). Retrieved from CriEnglish.com: http://english.cri.cn/about/history.htm.

Hoffman, D. S. (1997). *High tech hate: Extremist use of the Internet.* New York: Anti-Defamation League.

Hope, K. R. (1985). Electoral politics and political development in post-independence Guyana. *Electoral studies, 4*(1), 57–68. doi:10.1016/0261-3794(85)90031-9.

Horten, G. (2003). *Radio goes to war: The cultural politics of propaganda during World War II.* Berkeley, CA: University of California Press.

Horton, D., & Wohl, R. R. (1956). Mass communication and para-social interaction: Observations on intimacy at a distance. *Psychiatry, 19*(3), 215–229.

House Committee on Naval Affairs. (1938). *Hearings authorizing the secretary of the navy to construct and maintain a government radio broadcasting station.* Washington, DC: Government Printing Office.

House of Representatives Committee on the Judiciary. (2004). *Internet streaming of radio broadcasts: Balancing the interests of sound recording copyright owners with those of broadcasters—Hearing before the Subcommittee on Courts, the Internet, and Intellectual Property of the Committee on the Judiciary.* Washington, DC: Government Printing Office.

Hudis, M. A. (1997, January 27). Navajo Elvis croons, KTNN fans swoon. *Mediaweek, 7*(4), 47.

Huesca, R. (1995). A procedural view of participatory communication: Lessons from Bolivian tin miners' radio. *Media, Culture & Society, 17*(1), 101–119. doi:10.1177/ 016344395017001007.

Humana, V. (1923, February 24). Radio fans. *The North-China Herald and Supreme Court & Consular Gazette*, p. 528.

Hundreds in city hear Europe's radio. (1924, November 26). *New York Times*, pp. 1, 2.

Impressive veri from Georgetown, Demerara. (1938, May 1). *All-Wave Radio*, p. 243.

In foreign lands: The inventor of wireless telegraphy. (1898, January 2). *New York Times*, p. 19.

India's new network of radio broadcasting. (1938, April 1). *Radio Craft*, pp. 670, 698.

Inklineglobal. (2001, June 25). Retrieved from The Internet Archive: https://archiv e.org/details/tucows_193515_WinFM.

International radio: 400 of Egypt's villages to get radios. (1938, January 12). *Variety, 129*(5), 35.

International radio: Coughlin an international "perhaps." (1937, November 17). *Variety, 128*(10), 43.

International radio: Coughlin's invitation. (1939, January 18). *Variety, 133*(6), 29.

International: Anti-commie programs aimed at Czech people launch NCFE station. (1951, May 2). *Variety, 182*(8), 11.

International: India's radio in slow progress. (1949, January 26). *Variety, 173*(7), 13.

Isaksen, J. L. (2012). Resistive radio: African Americans' evolving portrayal and participation from broadcasting to narrowcasting. *The Journal of Popular Culture, 45*(4), 749–768. doi:10.1111/j.1540-5931.2012.00956.x.

Italy doesn't itch for war: Duce to U.S. (1931, January 2). *Chicago Daily Tribune*, p. 1.

Jack Cooper is announcer over WSBC. (1930, January 25). *The Chicago Defender*, p. 6.

Jack Cooper proves he's the perfect host. (1930, August 9). *The Pittsburgh Courier*, p. 16.

Jacso, P. (1996, April 1). The players in the real-time audio business. *Information Today*, p. 45.

James, E. K. (1922, November 1). Las maravillas del radio. *La Nueva Democracia*, pp. 27, 31.

Jayaram, N. (2000). The dynamics of language in Indian diaspora: The case of Bhojpuri/Hindi in Trinidad. *Sociological Bulletin, 49*(1), 41–62.

Jeansonne, G. (2012). The priest and the president: Father Coughlin, FDR, and 1930s America. *The Midwest Quarterly, 53*(4), 359–373.

Jenkins, A. F. (2008). Fighting hate speech: Hispanic organizations and responsible media take aim. *The Hispanic Outlook in Higher Education, 19*(3), 34–36.

Jenkins, C. F. (1925). *Vision by radio, radio photographs, radio photograms.* Washington, DC: National Capital Press Inc.

Johnson, P. (2005). KJLH-FM, South Central's hub to the world: Black radio's emergence on the Internet. *Convergence, 11*(1), 26–47. doi:10.1177/135485650501100104.

Jollife, A. (1925, January 22). Zulu war dance heard on radio. *New York Herald-Tribune*, p. 18.

Jungle broadcasts. (1937, September 24). *World Radio*, p. 6.

Jungle radio in South America: The Terry-Holden expedition. (2011, November 25). Retrieved from Shortwave Central: http://mt-shortwave.blogspot.com/2011/11/jungle-radio-in-south-america-terry.html.

Kahlenberg, R. S. (1966). Negro radio. *Negro History Bulletin, 29*(6), 127–143.

Keith, M. C. (2009). Norman Corwin's one world flight: The found journal of radio's greatest writer. *Journal of Radio & Audio Media, 16*(1), 50–65. doi:10.1080/193765 20902847972.

Khan, A. (1947, September 9). Letter to Kamaluddin Mohammed. *Correspondence. January 1946–October 1948, Box 1(Folder 5).* St. Augustine, Trinidad & Tobago: Special Collections, West Indian: The Alma Jordan Library.

King George's voice may girdle the globe when he opens British exhibition April 23. (1924, April 23). *New York Times*, p. E1.

Kirch, M. S. (1968). Happy birthday, Deutsche Welle. *Die Unterrichtspraxis / Teaching German, 1*(2), 117–118.

Klapper, J. T. (1960). *The effects of mass communication.* New York, NY: The Free Press.

Knoll, S. (1968, June 5). Radio, in "format" era, much like newspaper without a front page. *Variety, 251*(3), 31.

Knopper, S. (1996, December 21). Ethnic radio's potent, growing niche. *Billboard, 108*(51), 77–78.

Koahnic Broadcast Corporation. (2016, January 1). *Stations*. Retrieved from Native Voice One: Native American radio network: http://www.nv1.org/stations-affiliates/.

König, W. (2003). Der volksempfänger und die radioindustrie. ein beitrag zum verhältnis von wirtschaft und politik im nationalsozialismus. *VSWG: Vierteljahrs-chrift Für Sozial-Und Wirtschaftsgeschichte, 90*(3), 269–289.

Krebs, A. (1979, October 28). Charles Coughlin, 30's "radio priest," dies. *The New York Times*, p. 44.

Krysko, M. A. (2007). "Gibberish" on the air: Foreign language radio and American broadcasting, 1920–1940. *Historical Journal of Film, Radio and Television, 27*(3), 333–355. doi:10.1080/01439680701443101.

Krysko, M. A. (2011). *American radio in China: International encounters with technology and communications, 1919–41*. London: Palgrave Macmillan UK. doi:10.1057/97802 30301931.

Kuhn, F. (2011). Internet radio flows: Between the local and the global. *The Radio Journal—International Studies in Broadcast and Audio Media, 9*(1), 35–49. doi:10.1386/rjao.9.1.35_1.

Lal, B. V. (1998). Understanding the Indian indenture experience. *South Asia: Journal of South Asian Studies, 21*(1), 215–237.

Landry, R. J. (1943). The impact of OWI on broadcasting. *The Public Opinion Quarterly, 7*(1), 111–115.

Lang, J., & Simon, A. (1942, July 4). Foreign language radio. *Radio Daily*, p. 77.

Language news programs. (1937, April 9). *Radio Daily, 1*(43), 1, 3.

Lasar, M. (2016). *Radio 2.0: Uploading the first broadcast medium*. Santa Barbara, CA: Praeger.

Lasswell, H. D. (1927). *Propaganda technique in the World War*. London: Kegan Paul, Trench, Trubner & Co.

Launch all Negro radio show. (1951, May 19). *The Atlanta Daily World*, p. 1.

Lavine, H., & Wechsler, J. (1940). *War propaganda and the United States*. New Haven, CT: Yale University Press/Institute for Propaganda Analysis.

Lent, J. A. (1982). Mass media and socialist governments in the Commonwealth Carib-bean. *Human Rights Quarterly, 4*(3), 371–390. doi:10.2307/762224.

Lescarboura, A. C. (1922). *Radio for everybody; being a popular guide to practical radio-phone*. New York, NY: Scientific American Publishing Company/Munn and Compa-ny.

Lewis, A. T. (1933, March 4). History of advertising turns a new page with modern radio. *The China Press*, p. 17.

Lewis, J. O. (1999). From West Indian federation to Caribbean economic community. *Social and Economic Studies, 48*(4), 3–19.

Lindell, I. V. (2006). Wireless before Marconi. In T. K. Sarkar, R. J. Mailloux, A. A. Oliner, M. Salazar-Palma & D. L. Sengupta (Eds.), *History of wireless* (pp. 247–266). Hoboken, NJ: John Wiley & Sons Inc.

Listen in. (1925, January 20). *The China Press*, p. 2.

Local broadcast experiments. (1928, June 27). *The Daily Argosy*, p. 4.

Local broadcasting. (1927, February 24). *The Daily Chronicle*, p. 4.

Local broadcasting service: Heeney-Tuney fight to be relayed. (1928, July 25). *The Daily Chronicle*, p. 7.

Lochte, R. H. (2000). Invention and innovation of early radio technology. *Journal of Radio Studies, 7*(1), 93–115.

Lochte, R. H. (2001). *Kentucky farmer invents wireless telephone! But was it radio?* Murray, KY: All About Wireless.

Lovell, S. (2013). Broadcasting Bolshevik: The radio voice of Soviet culture, 1920s–1950s. *Journal of Contemporary History, 48*(1), 78–97. doi:10.1177/00220094 12461817.

Lowenthal, D. (1972). Black power in the Caribbean. *Economic Geography, 48*(1), 116–134.

Ludlam, G. P. (1945, January 8). Radio participation in war information campaigns—1944. *Broadcasting, Broadcast Advertising, 28*(2), 18.

Luthra, H. R. (1986). *Indian broadcasting*. New Delhi: Publications Division, Ministry of Information and Broadcasting, Govt. of India.

Lwanda, J. (2014). Chattering classes: Radio, rhythm and resistance in "multi-party" Malawi, 1994–2014. *The Society of Malawi Journal, 67*(2), 19–33.

Mabweazara, H. M. (2013). "Pirate" radio, convergence and reception in Zimbabwe. *Telematics and Informatics, 30*(3), 232–241. doi:10.1016/j.tele.2012.02.007.

Machlup, F. (1962). *The production and distribution of knowledge in the United States*. Princeton, NJ: Princeton University Press.

Mackenzie, H. (1999). *The directory of the Armed Forces Radio Service series*. Westport, CT: Greenwood Press.

Malamud, C. (1993). *Internet town hall*. Alexandria, VA: Internet Multicasting Corporation.

Manley, M. (1987). *Up the down escalator; development and the international economy: A Jamaican case study*. Washington, DC: Howard University Press.

Mann, J. (1997, December 3). National perspective: U.N. hate-radio jamming would send wrong signal. *Los Angeles Times*, p. 5.

Marconi scores new triumphs: Sends message of president to king from Cape Cod to Poldhu Wales. (1903, January 20). *Chicago Daily Tribune*, p. 5.

Marconi wireless: Great fortunes to be founded in the new system. (1903, September 3). *The Advance*, p. 258.

Market reports to farmers by wireless. (1921, June 18). *The Wall Street Journal*, p. 3.

Marsot, A. L. (1985). *A short history of modern Egypt*. Cambridge: Cambridge University Press.

Martin, H. (2019, March 24). *Fury as Twitter and YouTube promote "Wales-based" neo-Nazi "Radio Aryan" station that defended New Zealand mosque gunman's hate-filled views*. Retrieved from Daily Mail News: https://www.dailymail.co.uk/news/article-6844317/Fury-Twitter-YouTube-promote-Wales-based-neo-Nazi-Radio-Aryan-station.html.

Martin, P. (1988). More notes on a regional news agency: The Caribbean News Agency (CANA) and an alternative route to audiences of an advanced country, Britain. *Gazette: The International Journal of Mass Communications Studies, 42*(2), 71–80. doi:10.1177/001654928804200201.

Mass communication priority. (1980, May 1). *Guyana Chronicle*, p. 6.

Matthews, P. H. (1940, May 4). In Caribbean. *New York Amsterdam News*, p. 8.

Mazzenga, M. (2008). Condemning the Nazis' "Kristallnacht": Father Maurice Sheehy, the National Catholic Welfare Conference, and the dissent of Father Charles Coughlin. *U.S. Catholic Historian, 26*(4), 71–87.

McCormick, R. R. (1953, October 11). Radio in retrospect. *Chicago Daily Tribune*, p. 24.

McLellan, D. (2004, January 9). Obituaries: Thomas Stockham, 70, digital audio pioneer. *Los Angeles Times*, p. B16.

McPhail, T. L. (1987). *Electronic colonialism: The future of international broadcasting and communication*. Newburry Park, CA: Sage Publications.

Media has vital role in the revolution. (1979, February 27). *Guyana Chronicle*, pp. 1–2.

Men going to diamond fields. (1927, January 7). *New Daily Chronicle*, p. 4.

Metzl, J. F. (1997). Rwandan genocide and the international law of radio jamming. *American Journal of International Law, 91*(4), 628–651. doi:10.2307/2998097.

Migliore, S., & DiPierro, A. E. (1999). *Italian lives, Cape Breton memories*. Sydney: UCCB Press.

Mitra, A. P. (1997). At the dawn of radio science. *Science and Culture, 63*(1), 6–8.

Mock Yen, A. (2002). *Rewind: My recollections of radio and broadcasting in Jamaica*. Kingston, Jamaica: Arawak Publications.

Moffett, C. (1899, June). Marconi's wireless telegraph: Messages sent at will through space—telegraphing without wires across the English Channel. *McClure's Magazine, 13*(2), 99–112.

Mohammed, S. N. (2001). Personal communication networks and the effects of an entertainment-education radio soap opera in Tanzania. *Journal of Health Communication, 6*(2), 137–154.

Mohammed, S. N. (2007). Cable modems. In H. Bidgoli, *The handbook of computer networks* (Vol. 2, pp. 221–229). Hoboken, NJ: John Wiley and Sons.

Mohammed, S. N. (2011). *Communication and the globalization of culture: Beyond tradition and borders.* Lanham, MD: Lexington Books.

Mohammed, S. N. (2012). Home virtual home? A case study of Trinidad and Tobago groups on Facebook. *NMEDIAC—The Journal of New Media and Culture, 8*(1). Retrieved from http://www.ibiblio.org/nmediac/summer2012/Articles/home_on_face book.html.

Mohammed, S. N. (2017). *Distant voices near: Historical globalizations and Indian radio in Trinidad and Tobago.* Mona, Jamaica: University of the West Indies Press.

Mohammed, S. N., & Thombre, A. (2017). An investigation of user comments on Facebook pages of Trinidad and Tobago's Indian music format radio stations. *Journal of Radio and Audio Media,* forthcoming.

Molisch, A. F. (2011). *Wireless communications* (2nd ed.). Chichester, West Sussex, UK: Wiley.

"Monitor." (1949, October 1). Around the broadcast bands. *Short Wave News, 4*(10), 258–261.

Morecroft, J. H. (1921). *Principles of radio communication.* New York, NY: John Wiley and Sons Inc.

Morrison, C. A. (1941, November 15). Wanted: Listeners to Jamaica radio. *Movie Radio Guide,* p. 35.

Morrison, C. A. (1942, December 12). Short waves: Melbourne short-wave station describes troop landings at Buna. *Movie Radio Guide,* p. 12.

Mosime, S. T., & Mhlanga, B. (2016). Historical entanglements, conflicting agendas and visions: Radio Botswana and the making of a national radio station. *Journal of African Media Studies, 8*(1), 55–73. doi:10.1386/jams.8.1.55_1.

Moyo, L. (2012). Participation, citizenship, and pirate radio as empowerment: The case of Radio Dialogue in Zimbabwe. *International Journal of Communication, 6*(1), 484–500.

Mugglebee, R. (1933). *Father Coughlin: The radio priest, of the Shrine of the Little Flower.* Garden City, New York: Garden City Publishing Co., Inc.

Mukherjee, D. C., & Sen, D. (2007). A tribute to Sir Jagadish Chandra Bose (1858–1937). *Photosynthesis Research, 91*(1), 1–10. doi:10.1007/s11120-006-9084-6.

Munro-Smith, H. (2004). Performing gender in the Trinidad calypso. *Latin American Music Review, 25*(1), 32–56. doi:10.1353/lat.2004.0008. doi:10.1353/lat.2004.0008.

Music by wireless. (1921, May 7). *Billboard, 33*(19), 7.

Myles, J. F. (2000). Carnival radio: Soca-calypso music and Afro-Caribbean voice in a restricted service license station in Manchester. *Journal of Communication Inquiry, 24*(1), 87–112. doi:10.1177/0196859900024001006.

Naps' many famous faces. (2014, January 5). *Trinidad Express Newspaper,* p. 8. Retrieved from http://www.trinidadexpress.com/news/Naps-many-famous-faces-23 8802861.html#main.

National Broadcasting Company. (1944). *The fourth chime.* New York, NY: National Broadcasting Company.

National Council of Indian Culture. (2015, September 12). *Images of indentureship.* Retrieved from National Council of Indian Culture: Trinidad and Tobago: http:// www.ncictt.com/resources/indentureship.

National media must operate in the public interest. (1983, February 20). *Sunday Chronicle,* p. 10.

National negro radio hour. (1938, January 12). *Atlanta Daily World,* p. 6.

NBC soldier letters disclose interest in top programs sent around the world. (1942, July 20). *Broadcasting*, p. 47.

Ndlovu, M. (2018, November 1). *Zim to license community radio stations.* Retrieved from https:// bulawayo24.com/index-id-news-sc-national-byo-148689.html.

Negroponte, N. (1995). *Being digital.* New York, NY: Vintage.

Neigh, J. (2013). The transnational frequency of radio connectivity in Langston Hughes's 1940s poetics. *Modernism/Modernity, 20*(2), 265–285.

Neptune, H. R. (2007). *Caliban and the Yankees: Trinidad and the United States occupation.* Chapel Hill, NC: University of North Carolina Press.

New Mexican border stations threaten interference in U.S. (1931, November 1). *Broadcasting*, p. 10.

News in brief. (1939, October 06). *The Times*, p. 7.

News notes: From foreign lands. (1931, December 1). *Broadcasting*, p. 26.

Newton, D. (2008). Calling the West Indies: The BBC World Service and Caribbean voices. *Historical Journal of Film, Radio and Television, 28*(4), 489–497.

Nichols, J. S. (1984). When nobody listens: Assessing the political success of Radio Martí. *Communication Research, 11*(2), 281–304. doi:10.1177/009365084011002009.

Noriega, C. A., & Iribarren, F. J. (2013). Toward an empirical analysis of hate speech on commercial talk radio. *Harvard Journal of Hispanic Policy, 25*, 69–96.

Norman, H. (1998). The talk of the town: The saga of Bob Grant. *Journal of Popular Culture, 32*(1), 91–98.

Northrup, D. (1995). *Indentured labor in the age of imperialism, 1834–1922.* Cambridge: Cambridge University Press.

Norway experts see Russia links to Nordic extremist movement. (2018, February 13). *BBC Monitoring European*, p. 1.

Nuggets. (1897, August 14). *New York Times*, p. 4.

O'Connor, A. (1990). The miners' radio stations in Bolivia: A culture of resistance. *Journal of Communication, 40*(1), 102–110. doi:10.1111/j.1460-2466.1990.tb02254.x.

Odeven, E. (2005, November 9). Navajo radio picks up NFL broadcast. *Arizona Daily Sun.* Retrieved from http://azdailysun.com/navajo-radio-picks-up-nfl-broadcast/article_450fcfca-cfc4-5f57-b891-184147954d8e.html.

OECS radio news exchange. (2002, April 1). *Caribbean Update, 18*(3), 3.

Office of United States Chief of Counsel for Prosecution of Axis Criminality. (1946). *Nazi conspiracy and aggression volume.* Washington, DC: United States Government Printing Office.

Ogunyemi, O. (2006). The appeal of African broadcast web sites to African diaspora: A case study of the United Kingdom. *Journal of Black Studies, 36*(3), 334–352. doi:10.1177/0021934704273914.

One Caribbean Media. (2018, June 6). *Radio: The radio group.* Retrieved from https:// www.onecaribbeanmedia.net/media-group/radio/.

Onuzulike, U. (2014). "Discussing the Igbo language on the Igbo Internet radio: Explicating ethnolinguistic vitality. *Journal of African Media Studies, 6*(3), 285–298. doi:10.1386/jams.6.3.285_1.

OWI director to begin series of broadcasts. (1943, March 6). *New Journal and Guide*, p. 5.

OWI radio time seizure branded abuse of power. (1943, March 15). *Chicago Daily Tribune*, p. 3.

Palenchar, J. (2000, March 6). Tabletop radios deliver streaming Internet audio. *Twice, 15*(6), 27.

Papa, M. J., Singhal, A., & Papa, W. H. (2005). *Organizing for social change: A dialectic journey of theory and practice.* Thousand Oaks, CA: Sage Publications Inc.

Papa, M. J., Singhal, A., Law, S., Pant, S., Sood, S., Rogers, E. M. & Shefner-Rogers, C. L. (2000). Entertainment-education and social change: An analysis of parasocial interaction, social learning, collective efficacy, and paradoxical communication. *Journal of Communication, 50*(4), 31–55. doi:10.1111/j.1460-2466.2000.tb02862.x.

Park, R. E. (1922). *The immigrant press and its control.* New York and London: Harper & Brothers Publishers.

Patriotic garden party. (1927, January 22). *New Daily Chronicle*, p. 8.

Pauli, C. (2010). Killing the microphone: When broadcast freedom should yield to genocide prevention. *Alabama Law Review, 61*(4), 665–700.

Pfaff, K. V., & Toma, C. (2003). Audio quality of Internet radio systems. *Proceedings of the 2003 International Symposium on Signals, Circuits and Systems. I*, pp. 301–304. Bankok, Thailand: IEEE. doi:10.1109/SCS.2003.1227008.

Plans education by radio. (1922, April 22). *The Amaroc News*, p. 1.

Police of France will censor radio: Propaganda by enemies and market rigging are feared if air is uncontrolled. (1924, November 9). *New York Times*, p. 2.

Potter, S. J. (2008). Who listened when London called? Reactions to the BBC Empire Service in Canada, Australia and New Zealand, 1932–1939. *Historical Journal of Film, Radio and Television, 28*(4), 475–487. doi:10.1080/01439680802310282.

Potter, S. J. (2012). *Broadcasting empire: The BBC and the British world, 1922–1970*. Oxford, UK: Oxford University Press.

Poulter, J. (2018). LAUSD teacher Dara Bloom—sugar mama of British neo-Nazis. *Turning the Tide, 30*(3), 6–12.

Prayer, M. (1991). Italian fascist regime and nationalist India, 1921–45. *International Studies, 28*(3), 249–271. doi:10.1177/0020881791028003002.

Premdas, R. R. (1971). Guyana: Socialism and destabilization in the western hemisphere. *Caribbean Quarterly, 25*(3), 25–43.

Public indifferent, Coughlin quits. (1936, November 21). *The Daily News*, p. 1.

Radio. (1935, January 11). *The Daily Argosy*, p. 8.

Radio broadcasters look at taking business on-line. (2000, April 1). *Music Business International*, p. 31.

Radio broadcasts. (1936, January 16). *The Indian Listener*, pp. 88–126.

Radio Corporation of America. (1922, June 22). *Radio enters the home*. New York, NY: Radio Corporation of America.

Radio Demerara. (1960). *Radio Demerara in pictures*. Georgetown, Guyana: Radio Demerara.

Radio Dialogue falls on hard times. (2017, May 17). Retrieved from https://www.newzimbabwe.com/radio-dialogue-falls-on-hard-times/.

Radio Dialogue manager case today. (2013, March 5). Retrieved from https://www.newsday.co.zw/2013/03/radio-dialogue-manager-case-today/.

Radio Free Europe begins. (1950, July 17). *Broadcasting, Telecasting, 39*(3), 66.

Radio heard 7,880 miles. (1924, February 13). *New York Times*, p. 20.

Radio highlights and headlines: 1942. (1943, January 1). *Broadcasting, Broadcast Advertising, 24*(5), 56, 58.

Radio in Egypt. (1937, December 7). *The Times of India*, p. 19.

Radio in the Cold War. (1954, June 1). *World Today, 10*(6), 245–254.

Radio notes and gossip. (1924, February 17). *New York Times*, p. 14 xx.

Radio notes: Short-wave transmission. (1924, May). *Scientific American*, p. 357. doi:10.1038/scientificamerican0524-356.

Radio plans of Russia. (1947, May 19). *Broadcasting, Telecasting, 32*(20), 32.

Radio politics. (1939, September 29). *The Times*, p. 6.

Radio preached hate. (2017, July 28). *The Times*, p. 4.

Radio programme. (1940, May 4). *Trinidad Guardian*, p. 4.

Radio progams. (1943, February 4). *C.B.I. Roundup*, p. 2.

Radio suspected of too high morals. (1927, January 14). *New Daily Chronicle*, p. 5.

Radio weapon in the Middle East. (1955, May 17). *The Times*, p. 8.

Radio: Egypt station awaits 1941 for $385,000. (1940, March 20). *Variety, 133*(2), 34.

Radio-broadcasting in China. (1925, January 9). *The China Press*, p. 10.

"Radiocasting" is adopted by electrical men. (1924, June 22). *Chicago Daily Tribune*, p. E7.

Radiocasting of heat for homes now forecast: It's only a matter of few years. (1926, January 24). *Chicago Daily Tribune*, p. 3.

Radio's short waves. (1937, January 17). *The New York Times*, p. 164/12X.

Radio-television: RAB sets up division for Spanish stations. (1966, October 12). *Variety, 244*(8), 56.

Ralli, T. (2005, August 30). Religion on demand: iPod helps the busy faithful catch up. *International Herald Tribune*, p. 15.

Ramasastri, J. (1959). Nationalization of broadcasting in India. *Indian Economic Journal, 7*(1), 62–73.

Rangaswamy, P. (2000). *Namasté America: Indian immigrants in an American metropolis.* State College, PA: Penn State University Press.

Rao, B. (1986). All India Radio: The new challenges. *International Communication Gazette, 38*(1–3), 101–113. doi:10.1177/001654928603800108.

Ratti, A. D. (1931). Wireless message of His Holiness Pope Pius XI broadcast to the whole world. Pamphlet. London: The Catholic Truth Society.

Reception in British Guiana. (1933, November 1). *Radio News*, p. 314.

Reception of CBS relay program reported good. (1932, March 1). *Broadcasting*, p. 15.

Reed, J. (1945, August 1). Howdy folks. *WIBW Round-up, 5*, 11. Retrieved from http://www.americanradiohistory.com/Archive-Station-Albums/WIBW/WIBW-1945-08.pdf.

Regal, B. (2005). *Radio: The life story of a technology.* Westport, CT: ABC-CLIO.

Remenih, A. (1951, March 18). Radio Moscow's pitchmen prove simply boring. *Chicago Daily Tribune*, p. A10.

Report of the British Broadcasting Committee. (1923, October 13). *Nature*, pp. 558–560.

Rhinegold, H. (1993). *The virtual community: Homesteading on the electronic frontier.* Reading, MA: Addison-Wesley Publishing Company.

Rhodes, A. (1976). *Propaganda: The art of persuasion—World War II.* (V. Margolin, Ed.) New York, NY: Chelsea House Publishers.

Rixon, P. (2015). Radio and popular journalism in Britain: Early radio critics and radio criticism. *Radio Journal: International Studies in Broadcast and Audio Media, 13*(1–2), 23–36. doi:10.1386/rjao.13.1-2.23_1.

Rodney, W. (1973). *How Europe underdeveloped Africa.* Dar-es-Salaam, Tanzania: Tanzania Publishing House.

Rogers, E. M. (1976). Communication and development: The passing of the dominant paradigm. *Communication Research, 3*(2), 213–240. doi:10.1177/009365027600300207.

Rogers, E. M. (2002). Intermedia processes and powerful media effects. In J. Bryant & D. Zillman (Eds.), *Media effects: Advances in theory and research* (pp. 199–214). Mawah, NJ: Lawrence Erlbaum Associates Inc.

Roopnarine, L. (2006). *Indo-Caribbean indenture: Resistance and accommodation.* St. Augustine, Trinidad & Tobago: University Press of the West Indies.

Rue, L. (1942, June 13). Hitler steps up political drive in Middle East: Timed with Crimean, Libyan thrusts. *Chicago Daily Tribune*, p. 5.

Russia: Russian embassy in Norway: Allegations that Russia supports ultra-radicals are impudent. (2018, February 9). *Asia News Monitor*, p. 1.

Rzepka, A. (2011). Various definitions of globalisation. *International Journal of Arts & Sciences, 4*(13), 453–461.

Saerchinger, C. (1938). *Hello America! Radio adventures in Europe.* Boston: Houghton Mifflin Company.

Said, E. W. (1978). *Orientalism.* New York: Vintage Books.

Samaroo, B. (1987). The Indian connection: The influence of Indian thought and ideas on East Indians in the Caribbean. In J. I. Singh (Ed.), *Indians in the Caribbean* (pp. 25–50). London: Oriental University Press.

San Francisco: Concert by radio. (1920, October 29). *Variety, 60*(10), 20.

Sandberg, J. (1994, May 4). Live from Vegas . . . it's talk radio, TV, rock on the Internet. *Wall Street Journal*, p. B9.

Sanders, R. (1978). *Broadcasting in Guyana.* London: Routledge & Kegan Paul Ltd.

Sarnoff, D. (1939). *Principles and practices of network radio broadcasting.* New York, NY: RCA Institutes Technical Press.

Satia, P. (2010). War, wireless, and empire: Marconi and the British warfare state, 1896–1903. *Technology and Culture, 51*(4), 829–853.

Says radio device ends "listening in." (1924, July 16). *New York Times*, p. 20.

Scales, R. P. (2010). Subversive sound: Transnational radio, Arabic recordings, and the dangers of listening in French colonial Algeria, 1934–1939. *Comparative Studies in Society and History, 52*(2), 384–417.

Scales, R. P. (2013). Métissage on the airwaves: Toward a cultural history of broadcasting in French colonial Algeria, 1930–1936. *Media History, 19*(3), 305–321. doi:10.1080/13688804.2013.817837.

Scannell, P., & Cardiff, D. (1991). *A social history of British broadcasting*. Oxford, UK: Blackwell.

Schiappa, E., Gregg, P. B., & Hewes, D. E. (2005). The parasocial contact hypothesis. *Communication Monographs, 72*(1), 92–115.

Schiller, H. I. (1976). *Communications and cultural domination*. Armonk, NY: M. E. Sharpe.

Schneider, J. (2014, March 1). When CBS got serious about shortwave. *Radio World, 38*(6), 22.

Schramm, W. (1964). *Mass media and national development: The role of information in the developing countries*. Redwood City, CA: Stanford University Press.

Schubert, P. (1928). *The electric word: The rise of radio*. New York, NY: The Macmillan Company.

Schulberg. (2019, February 7). *The neo-Nazi podcaster next door*. Retrieved from https://www.huffpost.com/entry/grandpa-lampshade-neo-nazi-white-supremacist_n_5c5b12f9e4b00187b557717a.

Schulze, J. (1999, November 2). Net radio tipped to explode. *The Age: Business*, p. 1.

Searle, C. (1991). The Muslimeen insurrection in Trinidad. *Race & Class, 33*(2), 29–43.

Senate Committee on Government Operations, Permanent Subcommittee on Investigations. (1954). *Voice of America*. Washington, DC: United States Government Printing Office.

Seneviratne, K. (2011). Community radio via public service broadcasting: The Kothmale model. *Journal of Radio & Audio Media, 18*(1), 129–138. doi:10.1080/19376529.2011.562830.

Sessions, J. E. (2011). *By sword and plow: France and the conquest of Algeria*. Ithaca: Cornell University Press.

Settel, A. (1938, February 9). "A cup of coffee and the radio" sets new tempo of night life in Egypt: Oldtimers no like. *Variety, 129*(9), 37.

Shanghai and Osaka talk on two-way radio: Successful tests are carried out. (1925, March 4). *The China Press*, p. 1.

Shanghai radio assocn. (1923, February 24). *North-China Herald and Supreme Court & Consular Gazette*, p. 25.

Shenfield, R. (1983, May 23). Compact disc in radio infancy. *Broadcast*, p. A12.

Sid-Ahmed, A. A. (1984). *Mass media and development in Sudan*. Ann Arbor, MI: University Microfilms International.

Sidel, M. K. (1984). New world information order in action in Guyana. *Journalism Quarterly, 61*(3), 493–498, 639.

Sing Tao Chinese Radio. (2019, April 30). *Home: Sing Tao Chinese Radio*. Retrieved from http://www.chineseradio.com/main/.

Singhal, A., & Rogers, E. M. (1999). *Entertainment-education: A communication strategy for social change*. Mahwah, NJ: Erlbaum.

Singhal, A., & Rogers, E. M. (2001). *India's communication revolution*. New Delhi: Sage Publications India Pvt. Ltd.

Singhal, A., Rogers, E. M., & Mahajan, M. (1999). The gods are drinking milk! Word-of-mouth diffusion of a major news event in India. *Asian Journal of Communication, 9*(1), 86–107.

Smethers, J. S. (2016). Who needs a terrestrial signal? Internet audio streams make waves in two Kansas radio markets. *Journal of Radio & Audio Media, 23*(1), 20–35. doi:10.1080/19376529.2016.1155129.

Smith, B. L., & Cornette, M. L. (1998a). Electronic smoke signals: Native American radio in the United States. *Cultural Survival Quarterly, 22*(2), 28–35.

Smith, B. L., & Cornette, M. L. (1998b). Eyapaha for today: American Indian radio in the Dakotas. *Journal of Radio Studies, 5*(2), 19–30. doi:10.1080/19376529809384543.

Smulyan, S. (1994). *Selling radio: The commercialization of American broadcasting 1920–1934.* Washington, DC: Smithsonian Institute Press.

Somerville, K. (2012). *Radio propaganda and the broadcasting of hatred.* Hampshire, UK: Palgrave Macmillan.

Spotlight: Vatican Radio goes digital. (2004, April 26). *Satellite News, 27*(17), 1.

St. Hilaire, A. (2011). *Kwéyòl in postcolonial Saint Lucia: Globalization, language planning, and national development.* Amsterdam, The Netherlands: John Benjamins Publishing Company.

Staley, D. J. (1998). Digital technology and the mythologies of globalization. *Bulletin of Science, Technology & Society, 18*(6), 421–425.

Station "V.P.3.B.G." on the air. (1935, February 13). *The Daily Argosy*, p. 4.

Station list. (1937, June 1). *Radio News*, p. 729.

Station notes. (1932, March 1). *Broadcasting*, p. 21.

Steigler, Z. (2014). Michael Savage and the political transformation of shock radio. *Journal of Radio & Audio Media, 21*(2), 230–246. doi:10.1080/19376529.2014.950147.

Sterling, H., & Kittross, J. M. (2002). *Stay tuned: A history of American broadcasting* (3rd ed.). Mahwah, N.J: Lawrence Erlbaum Associates.

Stone, E. W. (1919). *Elements of radiotelegraphy.* New York, NY: D. Van Nostrand Company.

Storr, J. M. (2011). The disintegration of the state model in the English speaking Caribbean: Restructuring and redefining public service broadcasting. *International Communication Gazette, 73*(7), 553–572. doi:10.1177/1748048511417155.

Straus, S. (2007). What is the relationship between hate radio and violence? Rethinking Rwanda's "Radio Machete." *Politics & Society, 35*(4), 609–637. doi:10.1177/003232 9207308181.

Stuart, F. (1963, April 1). What ever happened to radio? Languishing networks, burgeoning independents. *Challenge, 11*(7), 4–7.

Sudama, T. (1979). The model of the plantation economy: The case of Trinidad and Tobago. *Latin American Perspectives, 6*(1), 65–83. doi:10.1177/0094582X7900600104.

Svenkerud, P. J., Rao, N., & Rogers, E. M. (1998). Mass media effects through interpersonal communication: The role of "Twende na Wakati" on the adoption of HIV/AIDS prevention in Tanzania. In W. N. Elwood (Ed.), *Power in the blood: A handbook on aids, politics, and communication* (pp. 243–253). Mawah, NJ: Lawrence Erlbaum.

Swanson, D. J. (2010). Preaching, prosperity, and product sales: A profile of on-demand digital audio offerings of Christian renewalist ministries. *International Journal of Listening, 24*(2), 106–124.

Sykes, M. (1984, May). The compact disc, player and system. *Broadcast and Power, 30*(5), 367. doi:10.1049/ep.1984.0202.

Tacchi, J. (2000). The need for radio theory in the digital age. *International Journal of Cultural Studies, 3*(2), 289–298.

Tanikella, L. (2009). Voices from home and abroad: New York City's Indo-Caribbean media. *International Journal of Cultural Studies, 12*(2), 167–185. doi:10.1177/13678779 08099498.

Telecommunications Authority of Trinidad and Tobago. (2016). *Annual market report 2015, telecommunications and broadcasting sector.* Barataria, Trinidad & Tobago: Telecommunications Authority of Trinidad and Tobago.

Telegraphy without wires: Prof. Slaby of Berlin exchanges messages at twenty-one kilometers. (1897, October 9). *New York Times*, p. 7.

The America Cup: Columbia and Shamrock. (1899, October 21). *Forest and Stream; A Journal of Outdoor Life, Travel, Nature Study, Shooting, Fishing, Yachting*, p. 336.

The British Broadcasting Corporation. (1928). *The British Broadcasting Corporation: First annual report, 1927*. London: His Majesty's Stationery Office.

The British Broadcasting Corporation. (1929). *The British Broadcasting Corporation. Second annual report 1928*. London: His Majesty's Stationery Office.

The British Broadcasting Corporation. (1932). *The British Broadcasting Corporation: Fifth annual report, 1931*. London: His Majesty's Stationery Office.

The British Broadcasting Corporation. (1947). *The British Broadcasting Corporation: Annual report and accounts for the year 1945–46*. London: Her Majesty's Stationery Office.

The British Broadcasting Corporation. (1966). *British Broadcasting Corporation: Annual report and accounts for the year 1964–65*. London: Her Majesty's Stationery Office.

The enter*active file: Merchants & marketing: New RealAudio is FM quality. (1995, November 18). *Billboard, 107*(46), 88.

The finance and future of broadcasting. (1926, March 13). *The Economist*, p. 504.

The Indian war effort. (1939, December 29). *The Times*, p. 7.

The opening match: M.C.C. v. British Guiana. (1935, February 5). *The Daily Argosy*, p. 8.

The Radio Staff of the Detroit News. (1922). *"WWJ—the Detroit News"; the history of radiophone broadcasting by the earliest and foremost of newspaper stations; together with information on radio for amateur and expert*. Detroit, MI: The Evening News Association.

The University of the West Indies. (2013, November 20). *Spoken word open day at University Archives, UWI, Mona*. Retrieved from https://uwiarchives.wordpress.com/tag/radio-education-unit/.

The World Bank Group. (2010). *Data, Trinidad and Tobago*. Retrieved from http://data.worldbank.org/country/trinidad-and-tobago.

This is station 2LO, London: Britishers listen eagerly to their own broadcast programs. (1922, October 1). *The Wireless Age, 20*(1), 27–29.

Thompson, M. E., Gómez, K. A., & Toro, M. S. (2005). Women's alternative Internet radio and feminist interactive communications: Internet audience perceptions of Feminist International Radio Endeavour (FIRE). *Feminist Media Studies, 5*(2), 215–236. doi:10.1080/14680770500124306.

Tignor, R. L. (2010). *Egypt: A short history*. Princeton: Princeton University Press.

Ting, C., & Wildman, S. (2003). The economics of Internet radio. *The Radio Conference Proceedings* (pp. 27–44). Madison, WI: University of Wisconsin-Madison.

Topics of the times. (1897, July 19). *New York Times*, p. 4.

Torosyan, G., & Munro, C. (2010). Earwitness testimony: Applying listener perspectives to developing a working concept of "localism" in broadcast radio. *Journal of Radio & Audio Media, 17*(1), 33–47. doi:10.1080/19376521003738037.

Toscano, A. (2012). *Marconi's wireless and the rhetoric of a new technology*. The Netherlands: Springer.

Towers, W. K. (1917). *Masters of space: Morse and the telegraph; Thompson and the cable; Bell and the telephone; Marconi and the wireless telegraph; Carty and the wireless telephone*. New York: Harper and Bros.

Travel and communication. (1924, September 1). *Current Opinion*, pp. 380–384.

trickleout.net. (2015, January 31). *Radio Dialogue*. Retrieved from https://www.trickleout.net/index.php/directoryofenterprises/Zimbabwe_/radio-dialogue.

Trinidad acts to keep control of mass media. (1968, December 30). *Ottawa Citizen*, p. 4.

Trinidad broadcast test success. (1935, January 15). *The Daily Argosy*, p. 8.

Trinidad station goes commercial. (1947, August 19). *Radio Daily*, pp. 1, 2.

Trinidad station: A description and the official report of successful working. (1915, April 1). *The Wireless World*, pp. 13–14.

Trinidad's commercial station preems in August. (1947, June 25). *Variety*, p. 26.

US Congress Senate Committee on Interstate Commerce. (1935). *Report on communication companies (House report 1273, part III no. 4)*. Washington, DC: Government Printing Office.

US House of Representatives. (1944). *Study and investigation of the Federal Communications Commission*. Washington, DC: Government Printing Office.

United States Broadcasting Board of Governors Special Committee on the Future of Shortwave Broadcasting. (2014). *To be where the audience is: Report of the Special Committee on the Future of Shortwave Broadcasting*. Washington, DC: Broadcasting Board of Governors.

United States Bureau of the Budget. (1946). *The United States at war; development and administration of the war program by the federal government*. Washington DC: Government Printing Office.

United States Congress, House Committee on Merchant Marine, Radio, and Fisheries. (1934). *Radio broadcasting*. Washington DC: Government Printing Office.

United States Informational and Educational Exchange Act of 1948. (1948, January 27). *Public Laws Ch 35, 36*. Retrieved from http://www.state.gov/documents/organization/177574.pdf.

United States Senate Committee on Foreign Relations. (1983). *Radio broadcasting to Cuba: Hearings before the Committee on Foreign Relations*. Washington, DC: Government Printing Office.

Vatican radio: Church . . . and state. (1991, February 18). *Variety, 342*(6), 64.

Vaughan, P. W., Regis, A., & St. Catherine, E. (2000). Effects of an entertainment-education radio soap opera on family planning and HIV prevention in St. Lucia. *International Family Planning Perspectives, 26*(4), 148–157.

Ventura Free Press. (1932). *The empire of the air; the story of the exploitation of radio for private profit, with a plan for the reorganization of broadcasting*. Ventura, CA: Ventura Free Press.

Venugopal, A. (2001, July 31). The sounds of home. *Little India, 11*(7), 20.

Viles, P. (1993, March 27). Coming soon: Talk radio via PC. *Broadcast and Cable, 123*(12), 27–28.

Vitello, P. (2014, January 2). Bob Grant, a combative personality on New York talk radio, dies at 84. *New York Times*, p. B14.

Walker, J. (2001). *Rebels on the air: An alternative history of radio in America*. New York, NY: New York University Press.

Wallerstein, I. (1974). *The modern world-system I: Capitalist agriculture and the origins of the European world-economy in the sixteenth century*. New York, NY: Academic Press.

Walpole, N. C., Arkin, S., Eisele, F. R., Giddens, J., John, H. J., Matthews, A., & Morrissey, D. H. (1965). *U.S. Army area handbook for Algeria*. Washington, DC: Government Printing Office.

Waltham, T. (1993, January 6). Weekly radio show to be new service on Internet. *Bangkok Post: Post Database*, pp. 1, 3.

WDAF to broadcast piano lessons this summer. (1924, June 29). *New York Times*, p. xx15.

Webb, A. (2014). *London calling: Britain, the BBC World Serice and the Cold War*. London: Bloomsbury.

Wells, H. G. (1935). *The new America: The new world*. London: The Shenval Press.

Westinghouse to cover country with radio entertainment. (1921, December 10). *Electrical Review, 79*(24), 887.

W-G-N, Tribune radio, starts tonight: Mayor Dever, M'Cutcheon on debut program. (1924, March 29). *Chicago Daily Tribune*, p. 1.

Wheen, A. (2011). *Dot-dash to dot.com: How modern telecommunications evolved from the telegraph to the Internet* (1st ed.). New York: Springer. doi:10.1007/978-1-4419-6760-2.

Whitefield, C. T. (1922, May 1). A tropical island radiophone: Radio adventures among the Bahama islands. *Radio Broadcast*, pp. 68–70.

Williams, E. (1962). *History of the people of Trinidad and Tobago*. Port-of-Spain, Trinidad: PNM Publishing Company.

Wilson-Heath, C. (1986). *Broadcasting in Kenya: Policy and politics, 1928–1984.* Anna Arbor, MI: University Microfilms International.

Winston, B. (1998). *Media technology and society: A history from the telegraph to the Internet.* New York: Routledge.

Wireless in Egypt. (1926, December 01). *The Times,* p. 13.

Wireless telegraphy in Trinidad. (1913, October 1). *The Wireless World,* pp. 427–428.

Wood, J. (1992). *History of international broadcasting* (3rd ed., Vol. I). London, UK: The Institution of Engineering and Technology in association with The Science Museum, London.

Wood, J. (2000). *History of international broadcasting* (Vol. II). London, UK: The Institution of Engineering and Technology in association with The Science Museum, London.

Yates, R. F. (1924, April 9). What will happen to broadcasting? *Outlook,* p. 604.

Youngman, E. P. (1934). *Information circular: Mining laws of British Guiana.* Washington, DC: US Bureau of Mines.

Yu, S. S. (2015). The inevitably dialectic nature of ethnic media. *Global Media Journal—Canadian Edition, 8*(2), 133–140.

Zaffiro, J. J. (1984). *Broadcasting and political change in Zimbabwe, 1931–1984.* Ann Arbor, MI: University Microfilms International.

Zambrano, W. R. (2018). La radio comercial en colombia. el nuevo panorama digital de la comunicación y del periodismo. *Signo y Pensamiento, 37*(72), 106–126. doi:10.111 44/Javeriana.syp37-72.rccn.

Zhangazha, T. (2018, November 3). *Strategy changes needed to push Zim media forward.* Retrieved from https://www.newsday.co.zw/2018/11/strategy-changes-needed-to-push-zim-media-forward/.

Index

196

Index

About the Author

Shaheed Nick Mohammed, PhD, is Associate Professor of Communications at the Altoona campus of the Pennsylvania State University. He is a former television journalist and health communicator who has worked and conducted research throughout the Caribbean, in Africa and in the Middle East. His publications include books and journal articles on the evolution of communications technologies in the context of global, historical and cultural forces.

Lightning Source UK Ltd.
Milton Keynes UK
UKHW012238081019
351248UK00001B/24/P